Farming in Britain

Frank Huggett

A & C Black London

Black's Junior Reference Books

General Editor R. J. Unstead

1	Houses	18	Costume
2	Fish and the sea	19	The story of the post
3	Travel by road	21	Castles
4	The story of aircraft	23	Coins and tokens
6	Mining coal	24	Fairs and circuses
7	The story of the theatre	25	Victorian children
8	Stars and space	26	Food and cooking
9	Monasteries	27	Wartime children 1939–1945
10	Travel by sea	28	When the West was wild
12	Musical instruments	29	Wash and brush up
13	Arms and armour	31	Canal people
14	Farming in Britain	32	The story of oil
15	Law and order	33	Home sweet home
16	Bicycles	34	Outdoor games
17	Heraldry	35	Horses

Acknowledgements

The author and publishers are grateful to the following for permission to reproduce photographs:

Aerofilms Ltd 43; *Trustees of the British Museum (Natural History)* 53b; *British Tourist Authority* 7a; *Cheviot Sheep Society* 15c; *Cowper & Co* 19b; *Crown Copyright reserved* 7c, 8b, 13, 14, 23a, 25, 39a, 41a, 49, 53a, 54b, 55, 58a, 59, 61b, d & e, 63a; *Farmers Weekly* 1, 2, 3, 5a & b, 8a, 9, 11a & b, 12a, b & c, 16a, & b, 17a, 18, 19c, 26, 28a, 34a, 35, 36a, b & c, 39a 40a & b, 41b, 42a, 44a & b, 45, 46a & b, 47a, b & c, 48, 54a; *Imperial Chemical Industries Ltd* 8c, 16c, 20, 21, 27, 34b, 56a & b, 57, 60a & b, 61a & c, 62; *Mansell Collection* 39b; *Massey Ferguson* 10a, 29a & b, 30a & b, 31, 32b, 33, 38a & b; *Milk Marketing Board* 6b, 7b, 17b, 19a & d, 22, 23b & c, 24, 63b; *National Farmers' Union* 6a, 58b; *National Portrait Gallery, London* 28b; *Anthony Phelps* 50; *Poultry World* 51a & b, 52d & c; *Sport & General Press Agency* 52b; *John Topham Ltd* 15a; *University of Reading, Museum of English Rural Life* 32a, 37a, & b; *R P Watts Ltd* 15b.

Cover photograph Crown copyright

Published by A & C Black (Publishers) Limited, 35 Bedford Row,
London WC1R 4JH
ISBN 0-7136-1527-3
Reprinted with corrections 1982
Third edition 1975
© 1975 A & C Black (Publishers) Limited
Reprinted 1980, 1982

All rights reserved. No part of this publication may be reproduced, stored in a retrieval system, or transmitted, in any form or by any means, electronic, mechanical, photocopying, recording or otherwise, without the prior permission of A & C Black (Publishers) Limited.

Printed in Great Britain by BAS Printers Limited, Over Wallop, Hampshire

Tractor pulling an automatic potato planter

Contents

1	The pattern of farming	5
2	Sheep farming	11
3	Beef farming	17
4	Dairy farming	20
5	Mixed farming	25
6	The tractor driver	29
7	The hay harvest	33
8	Grain crops	36
9	The root harvest	40
10	Intensive farming	43
11	The pigman	44
12	The chicken farm	49
13	Fruit and vegetables	54
14	New plants for old	59
15	Helping the farmer	63
Index		64

Map showing the main mountainous regions in England, Scotland and Wales

1 The pattern of farming

A farmer is a man who makes his living from the land. Each chooses the kind of farming which is best suited to his own land. He is also influenced very greatly by the price that he is likely to receive for his produce. Many farmers with rich, flat land have *arable farms* which grow crops. Farmers with good pastures often keep milking cows on their *dairy farms*. Some farmers, with hilly land, keep sheep while other farmers specialise in pigs or poultry. And some farmers grow crops and keep animals as well. Their farms are called *mixed farms*.

The kind of farming depends on many things: on climate, slope, soil, and altitude. If we make a journey across Britain, we shall notice that as the scenery gradually changes, the kind of farming changes, too.

Let us start a journey by car from the south-west of England, on the slopes of Dartmoor in Devon. We can see at once that the land is high and the weather is too cold for anything to grow except rough grasses, heather, and scrubby bushes. As we cross the moors, we sometimes come across a wall made of a very hard stone called granite, but much of the land is divided only by small streams which, in winter, swell into raging torrents.

These lambs are only a few hours old

Sheep on Dartmoor. Notice the boulders among the grass

6　The pattern of farming

A mixed farm on the borders of Berkshire and Wiltshire

Dairy cows grazing in the gentle countryside of Dorset

There are no large towns and few villages on Dartmoor. The only sign of life is a lonely farm-house, built of stone, and sheltered in a valley.

On the moors, covered by a veil of chilling mist for half the year, roam a few wild ponies and flocks of Dartmoor sheep. The sheep have long shaggy coats which drag almost to the ground. Hill sheep farms, like these, provide some of the cloth we use and the lamb we eat.

As we drive towards the southern slopes of Exmoor we find that the scenery is less bleak. The land is greener and on the small farms, which are often worked by one family, bullocks are reared. These bullocks, such as the blood-red North Devons, are called store cattle. When they are a year or more old, they are sold to lowland farms, where they are grazed on the richer pasture-land until they are fat enough to be sent to market.

As we drive eastwards into Dorset, the countryside becomes gentler and the fields larger. This is pasture-land, where herds of dairy cows graze. Dairy farming areas like these give us our milk, and some of our butter and cheese, though some Dorset farmers now grow crops instead of keeping cattle.

The pattern of farming 7

Continuing our journey, we come to the New Forest, uncultivated except in a few clearings. Herds of wild ponies live here. They will trot right up to people looking for titbits of food.

Beyond this woodland region we come to the South Downs which stretch right across Sussex from Chichester to Eastbourne. Sheep—such as the Southdown—still graze on the close-cropped pastures of the Downs, as they have done for many centuries, but there are also many pigs, poultry, and dairy cows on the farms below the Downs. At the foot of the Downs, fields of all sizes, shapes, and colours are spread out like a patchwork quilt. Their colours come from the crimson of clover, the deep green of potato leaves, and the gold of ripening corn. This region has many mixed farms, with some arable land for growing crops, and some grassland for grazing cattle.

Ponies in the New Forest

Sheep grazing on close-cropped pasture

Tractor with plough and press. The press is used on some soils instead of a harrow

8 The pattern of farming

Tending tomato plants in a glasshouse in Sussex

Hops growing in Kent. The tall cone-shaped buildings in the background are oast-houses, where the hops were once dried (see p 58)

At Worthing, we see small farms which consist almost entirely of glasshouses, in which there are cucumbers and fat clusters of tomatoes ripening. This is an example of intensive farming, where only one or two special crops are grown.

When we drive north-east into Kent we see another example of intensive farming: the orchards of cherry, pear, or apple trees, and the hop-gardens with their neat lines of hops, used in making beer.

Travelling north-east we cross the Thames and drive right across Essex and Suffolk, where we see many market-gardens and mixed farms, until we reach East Anglia. We have reached some of the best arable land in the country—and some of that land was made by man.

Drained over the course of centuries from the swampy region surrounding the Wash, much of the Fenland lies below the level of the highest spring tides, so that water has to be pumped up from the fields to canals, which act as farm boundaries. In a few places the land is so low that people in boats can look down on the farmer working in his fields.

A combine harvester cutting wheat on an arable farm

Harvesting sugar beet in Norfolk

Everything is very neat and straight: there are only square fields, straight roads, and a flat horizon with a few poplar trees, a church spire or a farm-house. The arable farms of the Fens provide wheat and barley, potatoes, and other vegetables such as sugar beet.

By the end of this journey we have seen all the main kinds of farming practised in Britain: hill-sheep farming, stock-rearing, dairying, mixed, intensive, and arable farming. Of course, we could plan a different journey, say from the Welsh mountains into Hereford and Worcester and north into Cheshire; or we could drive from the Pennines down across the flat land of Humberside. But wherever we went we should notice similar changes in scenery and farming: sheep and cattle on the hills, bullocks on the lower slopes, dairy cows on the rich, damp pastures and intensive farming near the big towns.

Then, where the land is flatter and the climate is drier, we should find the grain crops—wheat and barley—being grown and harvested in big fields.

Galloway cattle on the hills of south-west Scotland

Scottish Blackface sheep—a very hardy breed

The west of Britain is generally hilly and it has a good deal of rain. Heavy rainfall is ideal for grass, which needs a cool, moist climate if it is not to become parched and brown. So most of the permanent grasslands, which are never ploughed up, are found in the west. Much of our livestock is kept on these pasturelands which curve up from Cornwall and Devon, through the Midlands, to Scotland.

Just imagine for a moment that you made a similar journey across England four or five hundred years ago. The picture would have been very different. In those days the farmland of a village consisted of two or three huge open fields, divided into narrow strips, each looking like a large allotment. Each villager cultivated a few strips in each field. He might also have some lean pigs, sheep or cows, which grazed with those of other villagers on waste land or commons.

By the beginning of the nineteenth century most of these large open fields had been divided up into separate farms, so that the farmer's land was then all in one piece and enclosed by hedges, ditches, walls, or other boundaries. Recently, many farmers have uprooted hedges to make bigger fields again, because they are easier to cultivate. Farming methods are changing all the time.

2 Sheep farming

It is winter. Around the hills of a farm in the north of Scotland a blizzard rages, piling up the snow into deep drifts.

Somewhere on the lower slopes of the hillside a sheep is buried in the snow. The mountain shepherd, accompanied by his collie dog, battles through the snow, searching for the sheep's tracks, listening for its bleating, and hoping that he will find it in time to save its life.

In winter, mountain shepherds have to make many expeditions on the hillsides, not only to rescue sheep which have strayed, but also to feed the main flocks. Normally the flock feeds on the heather and rough grasses which grow wild on the hills. But in winter, when the ground is covered with snow, the shepherd must take them hay (dried grass), or turnips from the precious store which he has grown at the foot of the hills during the short summer.

All through the winter the shepherd is very busy. Even when the heaviest snowfalls are past, the shepherd cannot rest, for in April or May the lambs are born. This is his busiest time—and light snow may still be falling. For a whole month many shepherds hardly go to bed.

The animals must be fed whatever the weather

A ewe licking clean her new-born lambs

The main sheep farming areas of Britain

Shearing on a farm on Exmoor

All this time the coats of the older sheep have been growing and, in late June or July, the sheep are gathered in from the hills so that their thick winter coats can be sheared. The clipper cuts the coat off all in one piece, using hand shears or a clipping machine. A fleece can weigh as much as 4½ kilos.

About a fortnight later the sheep are dipped. They are pushed into a large bath or made to swim through a long trough filled with a special disinfectant. This kills the little mites, shaped like beetles, which lay their eggs in the sheep's wool, causing a disease called sheep scab.

In August and September, some of the young male lambs which are not being kept for breeding are sold for fattening on lowland farms. Many hill farmers now keep some of the lambs and fatten them for sale on specially grown crops. These *wethers*, as they are called, will end up in the butcher's shop as lamb.

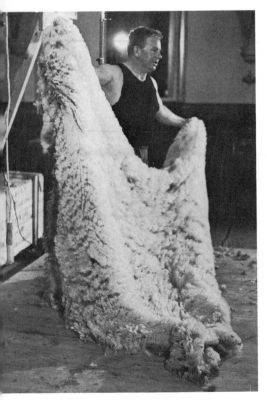

The world champion sheep shearer holding up a complete fleece

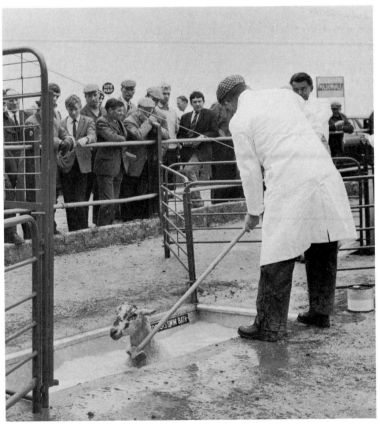

A demonstration of sheep dipping

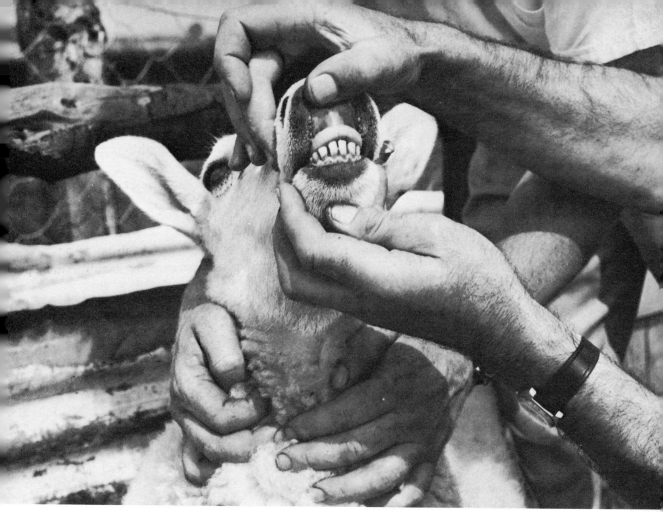

Inspecting a sheep's teeth to tell its age

At about the same time some of the older, female sheep (the ewes) are sold for mutton. Some of them are no longer capable of having lambs. Other ewes have to be sold because they can no longer eat. A sheep has front teeth only in its lower jaw, and these bite against the hard upper gum. The heather is so tough that after four or five winters the teeth are worn down to stumps. Because of this, a shepherd tells the age of a sheep by looking at its teeth.

Once the ewes and wether lambs have been sold, the shepherd has to prepare for the hard winter again. Every night of the autumn he drives the sheep up the hills so that in the morning they will start grazing at the top. This leaves the lower slopes for grazing in the winter, when the tops of the hills will be deep in snow.

More than two-thirds of the farms in Scotland are like this sheep farm. There are many similar farms in the central mountainous areas of Wales and on the Pennines.

Sheep farming

Some of these farms have six or seven thousand acres—more than thirty times as large as a medium-sized farm in the south. Every farm is divided into separate sections, each of which may have an area of two thousand acres. In charge of each section is a shepherd—and his dog. Without his dog the shepherd would not be able to control his flock. The most popular breed is the collie although others are also used. It takes two or three years to train a dog thoroughly.

The skill of these animals can be seen at the sheep-dog trials held in sheep farming areas each year. The dog has to drive five sheep for over 350 metres through narrow gateways, and herd them into a small pen. The shepherd signals to his collie by whistle, sign, or word. Hissing is a common signal to make a dog stop.

The dog runs, circles, and moves right or left at his master's command. By crouching down and fixing its eye on the sheep, the collie can hold the animal motionless. Much of the dog's skill consists in knowing when to stop still.

A shepherd and his sheep dog rounding up sheep on the moors

The collie will answer only to its own master's voice. Shepherds, therefore, never speak to another shepherd's dog. Some shepherds use two dogs. In Scotland they sometimes talk to one in English and to the other in Gaelic!

It is not only the dogs, however, who help the shepherd to control the flocks on the farm. Some of the older sheep are useful, too. The boundaries of the farm are not marked but, over the years, the older sheep come to know where they are. They will not let any of the flock stray outside or permit another farmer's sheep to come in. Usually a few old wethers are kept with the flock to act as 'policemen' in this way.

Of the thirty or more different breeds of sheep in Britain only mountain sheep could survive the harsh conditions on hill farms. Among the most common sheep in Scotland are the Cheviot and the Scottish Blackface, whose rough, hairy wool is used chiefly for making tweeds, carpets, and blankets.

In parts of the Midlands, Devon, and Kent, long-woolled sheep can be found. Their wool is so long and shaggy that it looks like a coat of straw. They provide wool for curtain materials and worsteds. The short-woolled sheep of the south, like the Southdown, give us dress lengths and other fine materials.

Scottish Blackface ewe

Cheviot ram

Lincolnshire ewe

16 Sheep farming

Wiltshire Horn ram

Border Leicester ewe

Oxford Down ewes

Although the Shetland is the smallest breed of sheep, its wool is the finest and the most valuable. In the Shetlands these sheep are not shorn, but their loose wool is picked off at intervals and used for making shawls. Another strange sheep is the Wiltshire Horn, which looks more like a goat. It has practically no wool. It is useful because its lambs are born early in the year and it can be mated with other breeds to produce early lambs.

Some sheep are also kept on the smooth, green pastures of the South Downs and the Cotswolds, but many farmers now keep cattle instead. In Scotland, too, because of rising costs and the difficulty of getting shepherds, some hill farms are now used for growing trees as well as keeping sheep. The pattern of farming in Britain is constantly changing.

Nevertheless, sheep still play a most important part in our farming, and there are about 30 000 000 sheep in Britain. Apart from poultry, more sheep are kept on our farms than any other animal.

3 Beef farming

Sheep farms are the largest in the country. Some of the smallest farms are found on the lower slopes of hills and mountains in Scotland, Wales, Northern Ireland, parts of the Midlands, and Devon. These stock-rearing farms produce bullocks, which are later fattened to provide meat.

Although the farms are not quite so isolated as hill sheep farms, they are often six to eight kilometres from the nearest town and the farmer leads a lonely life.

The calves are usually born in the spring or in the autumn. When they are about a fortnight old, they are taken away from their mothers and fed on milk substitute from a bucket. Gradually, they are weaned on to solid foods, sometimes when they are only six weeks old. They usually spend the first winter in an open or covered yard and in the spring they are turned out on to grass.

When they are a year to eighteen months old, they are sold as store (unfattened) cattle to other farmers. Farmers come from all over the country to some of the important store cattle sales. They take the cattle they have bought back to their own farms and fatten them by feeding them on hay, roots, and concentrated food in yards, or by letting them graze on rich pastures.

The main beef farming areas of Britain

An auction at Northampton cattle market

A newly-born calf

18 Beef farming

A bullock may eat as much as 68 kilos of grass in a day. Like a sheep, it can eat grass because it has two stomachs. But unlike sheep, which nibble at the grass, the bullock puts its tongue around a clump and draws it into its mouth. It does not chew the grass immediately, but swallows it so that it passes into the first stomach, called the *rumen*. It is stored there for a time and starts to ferment, just as the juice of grapes ferments when wine is being made.

Measuring the length of beef carcases

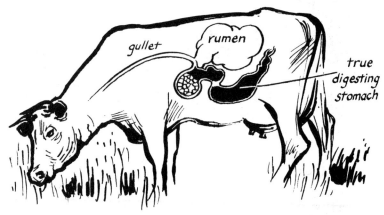
How a cow digests its food

When the bullock has eaten enough grass, it starts to bring the partly-fermented grass back into its mouth to chew the cud. Only after that does the grass go down into the real second stomach. Cows, sheep, goats, deer, and giraffes all digest their food in a similar way and are called *ruminants*.

Most of the cattle are two years old or less when they are sent to a fatstock market, where they are graded and weighed before being sent to the slaughter house.

Not many years ago, most cattle were not slaughtered until they were older. But people now want smaller joints of beef with little fat. To cater for this demand a different method of producing beef is being used increasingly. Under this intensive system, cattle are kept in covered yards and fed mainly on barley—as much as they can eat. In this way, they are ready for market in about a year. Friesians are most suitable for this.

Aberdeen Angus bull

Hereford bull

There are many different breeds of beef cattle. For its size, the Aberdeen Angus produces the greatest amount of meat. It is reared in the north-east of Scotland. Like the Galloway, which comes from the southern uplands of Scotland, it is a polled animal, that is, one without horns. Another important breed is the white-faced Hereford, bred on hill farms along the Welsh border. At one time it was used to draw the plough. In the last ten years many new breeds of cattle have been introduced on to British farms.

The pure white Charolais (top) was first imported from France in 1961. Later came the Simmental from Switzerland (bottom). Other breeds have been imported from Germany, Switzerland, the USA and Australia. These are all late maturing beasts which produce lean meat

4 Dairy farming

Richard Knight's father farms a hundred and fifty acres in Dorset. It is a dairy farm with a herd of a hundred black-and-white Friesian cows.

Richard has been helping his father on the farm ever since he left school. At five in the morning his alarm bell rings and he gets out of bed, rubbing his eyes, and starts to dress. Whenever he is milking he has to get up at the same time—winter and summer alike. After he has had a cup of tea, he goes off to fetch the cows from the bottom meadow.

Meanwhile, his father goes into the milking parlour next to the cow-yard beside their house. The parlour is a long, narrow building with a concrete floor; on one side are metal stalls, where the cows are milked.

Richard's father washes his hands and puts on his clean milking overalls. Then he gets a bucket of clean, warm water and a cloth and brush. As the first cows come into the yard Richard leads four of them into the stalls in the parlour.

His father puts a chain round each animal to keep it in its stall. Then he gives each one some cattle cake made from crushed seeds, rich in oil, such as coconuts or ground nuts.

The old method of milking cows by hand

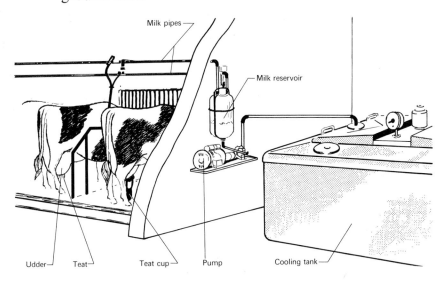

How a milking parlour works

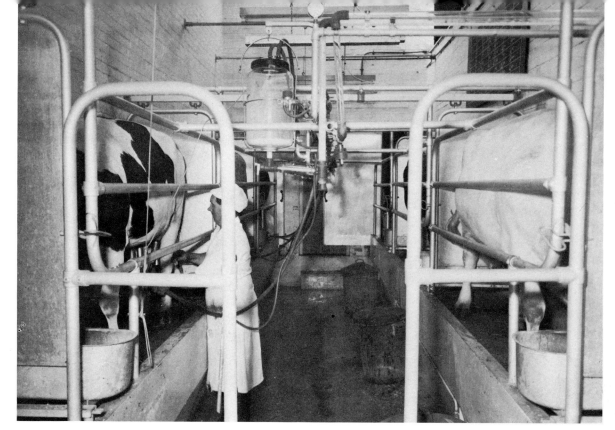

Inside a modern milking parlour

In summer the cows live mainly on grass, but they still need extra food if they are to give a good supply of milk. In winter, when there is little grass, they have to be fed on hay and cattle cake or root crops.

While his father feeds the animals, Richard gives each cow a good wash. First he gives it a brush to get rid of all the dirt. Then he washes its flanks, tail, and udder—the milk-bag which hangs down between the cow's back legs. This is a most important job. If it were not done, some dirt might get into the milk and make it unfit to drink.

Now the milking starts. Richard's father fixes the small rubber-lined cups of the milking machine on to the teats of the udder. The cups squeeze the teats and draw out the milk which is sent along a pipe to the milkroom, or dairy, at the end of the shed. There it flows into a bulk milk tank where it is cooled.

Meanwhile, Richard is taking the last drops of milk from the cows by milking each one by hand. In all it takes about five minutes to milk each animal. When all the animals have been milked the milking machine has to be rinsed with clean water.

The main dairy farming areas of Britain

Dairy farming

After Richard has taken the cows back to the fields and his father has fed the young calves, it is time for their own breakfast. While they are eating it, they hear the milk tanker arrive. The tanker sucks the milk out of the tank.

It is then taken to a creamery in the nearest town where most of the milk is heated for a short time to kill any germs—this is called *pasteurising*. Then it flows to a bottling plant, where modern machines fill two hundred bottles, or cartons, every minute.

After breakfast, work starts again on the farm. Richard and his father go back to the milking parlour, the floor of which has to be washed and scrubbed as carefully as in a hospital. The soiled straw is taken from the stalls and replaced by fresh straw. The buckets and parts of the milking machine are then scalded with steam, like a surgeon's instruments, to make certain that they are free from germs. The empty milk tank has to be washed and swilled out with clean water. Cleanliness is most important on a dairy farm if good milk is to be produced.

Hundreds of bottles being filled and sealed by machines at a milk bottling plant

Dairy farming

Friesian heifers about ten months old

Richard and his father then go to look at the heifers in another field. These are young cows which have not yet produced milk. A heifer starts to give milk as soon as it has its first calf at the age of eighteen to twenty months, and it will go on doing so for nine or ten months.

Only when it has had its second calf is the heifer called a cow. After the second calf is born, the mother will give milk for another nine months. Some cows live for twelve or thirteen years and have as many as ten calves, but most live less than half that time and produce only two or three calves.

Jersey cow

In the afternoon Richard and his father go back to the milking parlour again. The cows have to be milked twice a day. On average, cows give about 4400 litres of milk a year—that is, about 12 litres a day.

The largest quantity is given by the Friesians, the cows on Richard's farm, but their milk contains the least amount of butter-fat, from which butter is made. The creamiest milk comes from the smaller, Channel Island cows, the Jersey and the Guernsey, which are also the prettiest.

Guernsey cow

When foot-and-mouth disease infects a herd of cows, they must all be destroyed to stop the disease spreading. This picture was taken during the 1967–68 epidemic

Most dairy farms are found in Lancashire, West Yorkshire, Devon, Dorset, Somerset, Wiltshire, and south-west Scotland, but the dairy herd is also important on many mixed farms in other parts of the country.

There are always great risks in livestock farming. In spite of all the help of modern science and the farmers' own cleanliness and care, outbreaks of disease can never be entirely eliminated. In 1967–68 there was an epidemic of foot-and-mouth disease which was the worst ever recorded in Britain. In all, it is estimated that it cost farmers about £100 million.

Many herds which had been built up with such great care during the years had to be killed and burnt to prevent the disease spreading.

5 Mixed farming

So far we have visited only livestock farms in the north and the west of Britain. More and more farmers are becoming specialists of this kind. They concentrate on one kind of animal—cows for milking, calves to produce veal, or pigs and poultry, which we shall read about later. They keep much bigger herds of animals than farmers did in the past.

Even though they may grow some crops to feed their stock, their main interest is in just one kind of animal. Long before Britain joined the Common Market, this tendency to specialise was growing on many farms.

But we can find very different kinds of farm in the flatter, drier counties in the south and east. Much of the farmland is arable, and many of the farms are mixed. On a mixed farm, crops and livestock are both important. Several kinds of animals (such as cattle, poultry, pigs, and sheep) are usually kept and various kinds of crops (such as grain, potatoes, turnips, or kale) are grown.

Among some farmers, mixed farming is still popular because if bad weather ruins the farmer's harvest, he has a chance to make up his loss with the profits from his animals. Moreover, the grain and vegetables provide food for the animals, while the animals provide dung for the land to grow good crops. Many crops, like wheat, can be sold as cash crops to bring in ready money; while others, like kale, can be used as fodder, for feeding to animals.

The main areas of mixed farming in Britain

A mixed farm. Extra buildings have been put up over the years as they were needed

26 *Mixed farming*

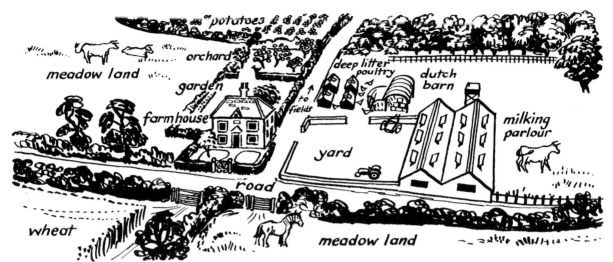

The variety of activities on a mixed farm

A fine crop of kale

Mixed farms vary greatly. On a typical mixed holding in south-east England, you may find a herd of 50 dairy cows, 100 pigs, and 500–1000 chickens.

Only a part of the 200 acres or so would be permanent grassland—perhaps three or four fields. An equal number of fields may be given over to temporary grasslands, called *leys*, which are also used for grazing or making hay. The rest may be growing barley, wheat, potatoes, kale or cabbage. On such a farm, two or three men may be employed, including a cowman and a tractor driver.

A mixed farm in East Anglia would be quite different. There, over two-thirds of the land would be arable with only a few pigs or cattle.

If you visited any of these farms regularly, you would notice that the crops grown in each field changed from year to year. One field which had been golden with ripening wheat the previous year would now be green with kale; another, which had been under grass would now be planted with oats.

If the same crop were always grown on the same land, the farmer would soon have trouble. The soil would become exhausted, and so the crops would be poor. Also, there is a danger of disease if the same crops are grown in the same field year after year.

Mixed farming 27

It was the Romans who first introduced into Britain a rotation of crops, extending over three years. This three-course rotation was:

1 Wheat, oats, or rye
2 Peas or beans
3 Fallow. The land was left bare and animals grazed on any wild vegetation.

This was a wasteful system because it left a third of the land idle for a whole year. Also, it gave little or no winter food for cattle and sheep. Many of them had to be killed in the autumn and their meat salted to preserve it for the winter months.

New farming methods were gradually introduced from the Continent in the seventeenth century. These involved the growing of turnips, which provided winter fodder for sheep and cattle, and the sowing of clover, which gave good grazing and winter hay.

These methods, which were first used in Flanders, are particularly suitable for light soils. More and more farmers on the light soils of Norfolk started to use them.

Cows grazing on ryegrass. The electric fence gives a weak shock, but the cows can only feel it if they touch the fence with their noses, so they could easily push the fence over if they wanted to

28 Mixed farming

Hay being stored as winter fodder for cattle

Viscount 'Turnip' Townshend, wearing his fine court dress

Then, in the eighteenth century, two rich landowners adopted these methods on their estates. These two men—Thomas Coke, Earl of Leicester, and Viscount Townshend, who became known as 'Turnip' Townshend—became famous for their successful methods. A new four-course rotation was used which became known as the Norfolk rotation. Under this, the order was:

1 Root crops
2 Barley, or spring oats
3 Clover
4 Wheat

The increase in the amount of hay, and the fodder provided by root crops, allowed more cattle to be kept through the winter. For the first time, people could enjoy eating fresh meat in winter instead of tough, salted meat. This new rotation of crops was not suitable for all types of land. But where it could be used, much more food was produced.

Even today, with fertilisers and other chemical aids, it is still necessary to practise some form of crop rotation. That is why you will rarely find a farmer sowing the same crop in the same field year after year.

6 The tractor driver

The tractor driver is one of the most important workers on a mixed or arable farm, as we shall see in the next few chapters.

Let us visit a farm in Essex to see how Bill, the tractor driver, prepares a field for sowing corn. It is a cold, autumn morning when Bill arrives at the tractor shed. First he fixes the plough at the back of his tractor. Then he climbs into his seat, starts the engine, and drives out of the yard to the field.

As Bill drives through the gate, there is very little space to spare on either side of the plough, but he has been driving for so many years that he barely has to glance over his shoulder as he goes through. It is a fine day for ploughing, as there has been no rain for several days. A field cannot be ploughed properly if the soil is wet and heavy.

The plough has a sharp blade, called the *coulter*, which cuts the soil vertically, and a *share*, which cuts horizontally, thus taking a slice out of the soil. This slice is turned over by the *mouldboard*, to form a ridge. Unlike the old horse-drawn plough, which made only one furrow, Bill's machine makes three, and some tractor ploughs can make as many as eight furrows at once.

The start of a day's work

How a plough cuts through the earth. The end of the field where the tractor turns is called the headland

Tractor with a three furrow plough

Tractor with cultivator

The depth of the cut can be varied too. Bill has adjusted the plough so that it will cut to a depth of fifteen centimetres only, since deep ploughing is not usually necessary for grain crops.

In Bill's hands driving a tractor looks easy, but it is not. In fact, the first time he ever ploughed, the furrow was as crooked as a corkscrew. At the end of each furrow a space is left for the tractor to turn. This is called the headland, and it is ploughed last. Some farmers have now abandoned ploughing in favour of direct drilling. The surface of the soil is scratched by a cultivator, and the seed is planted among the stubble of last year's crop. This method saves time and cuts the cost of tractor fuel.

When the whole field has been ploughed, Bill has to go over it twice more, with different tools. Corn will only grow well when there is a fine top layer, called the *tilth*. The second time he fixes a cultivator to his tractor. This has metal *tines*, or teeth, which dig into the plough ridges and break them down. Even this is not fine enough and he now uses another tool, called a disc harrow, which breaks the soil up even more. The discs cut into the soil and make a fine seed-bed, just as a rake does in the garden.

Using a disc harrow

Tractor pulling a seed-drill

If the ground were uneven he would use a chain harrow, which has tines fixed to a flexible chain, so that the soil is broken up evenly.

Now the ground is ready for sowing. This is done by a seed-drill, pulled by the tractor. The seed is carried in a long box, called the hopper, at the back of the drill. As a knife or disc opens a furrow in the ground, the seed is picked up by cups on the edge of a revolving wheel and dropped into the furrow. Bill again has to make sure that the furrows are straight, otherwise there would be bare patches in the field when the corn started to grow.

Even then, his work on this particular field is not finished. He harrows the ground again lightly to cover the seed and then rolls it to make it firm. And he will not see the full results of his work until next summer, when the corn will be ripening in the field—if the weather has been good.

32 The tractor driver

An automatic tractor

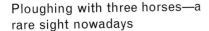

Ploughing with three horses—a rare sight nowadays

In a few parts of the country you will still see farmers ploughing with horses (particularly in hilly districts) but most farmers now use tractors. In the near future even the tractor driver may be unnecessary, as tests have already been made with an electronically controlled tractor that could do the ploughing by itself. In one test this automatic tractor hauled two trailers, stopped by itself at traffic lights, sounded its hooter, threw off bales of hay at chosen places, and stopped its engine when the work was done.

But the day when all ploughing on the farm is done by tractors like this has not yet arrived, and the tractor driver remains one of the key men on the farm.

Notice the rows of swaths which the mower has already cut to the left of the tractor

7 The hay harvest

Hay is usually harvested in June or July in the south. It is one of the most anxious times for the farmer as he relies on the hay to provide winter food for his cattle or other animals, and rain can spoil the whole crop, once it has been cut.

There is a ten acre field of hay to be harvested on the same farm in Essex. The meadow was shut early in the spring, so that the grass could grow tall and flower. Now that the seeds of the grass have just begun to lose their greenness the farmer knows that it is time to cut the hay. Bill, the tractor driver, fixes a mowing machine to his tractor and starts to cut the grass. By the end of the day all the grass is lying in long rows called *swaths*.

The hay must be dried as quickly as possible. This can be done in two main ways. Sometimes the grass is crimped, that is bent at five to seven centimetre intervals, so that the moisture is let out more quickly. This is normally done as the hay is cut, by a special machine called a grass crimper.

Turning hay with a side rake

An old fashioned haystack

Then the swaths must be frequently loosened (or *tedded*) and turned, so that the air can dry every part of them. One machine frequently used for this purpose is a finger wheel side rake which does both operations at once. The hay is tedded and turned until it is dry enough for storage. With these modern methods it is sometimes ready in two days instead of the five days or so that it used to take with older methods.

An automatic pick-up baler is then driven through the field. It gathers up the hay and automatically ties it into bales with wire or string. The bales are then collected by a buckrake or by a bale carrier attached to a tractor.

If they have to be carried far, they are usually loaded on to a trailer or a lorry, and taken to the farm-yard. There the bales are sometimes stacked and covered by a sheet of plastic to keep out the rain, or else they are stored in a Dutch barn, a huge shed with a roof but no sides. The bales are stacked in the barn with the help of an elevator.

The elevator has a long, wooden trough which can be raised or lowered to any height. Inside the trough is a series of spikes or prongs on an endless belt. When the bales are loaded on at the bottom, the prongs carry it to the top.

The hay harvest

On the most modern farms, the hay is finally dried by means of a large fan which blows warm air through ducts or through the slatted floor of the barn, in the same way as gas-fired central heating warms a room by circulating hot air through the little metal grilles in the walls.

Other farmers use another method of preserving grass. It is cut when it is still green and packed tightly into a tower called a *silo*. When the silo is full it is covered with a layer of earth and straw. The grass will keep fresh and moist until the winter, when it emerges as a sticky, yellowish-green substance, known as *silage*.

Whichever method of preserving grass is used, hay or silage is most important on any farm where livestock is kept; for without it there would be very little winter fodder for the animals.

Once the harvest has started it is important to gather all the hay while the weather is fine. In this field two tractors are at work—the one in the foreground is pulling a side rake, and the one in the background is baling the hay

8 Grain crops

By late summer the corn that was sown the previous autumn is ready to be harvested. Corn is the grain, or seed, of one of the cereal plants and with its many uses, both as food and fodder, it is the farmer's most important crop. Of the total area of ploughed land, more than half is used for growing corn.

Barley is the most popular crop. Although it grows best in the south and the south-east, it ripens so quickly that it can be sown as far north as the Orkneys and the Shetlands. In the last ten years, production of barley has doubled. It is used for making beer and for feeding livestock.

The second biggest cereal crop is wheat, which is grown mainly in England. It is a very useful crop because it is made not only into bread, but also into biscuits, breakfast foods and even macaroni.

Oats is the smallest of the three main crops. At one time it was widely grown because it was so useful for feeding to horses, but that is no longer necessary. Even in Scotland, where it was once a very popular crop, it has been replaced on many farms by barley. But porridge made from oatmeal is still the national breakfast dish of many Scots.

The main areas in Britain where grain crops are grown

Oats   Wheat

A field of barley

Victorian farm-workers harvesting with scythes

Very little rye is grown nowadays. In the past it was more widely sown, because it will grow well on almost any soil. The straw was ideal for thatching cottages and the grain could be made into a hard, black bread. But few houses are thatched nowadays and most of us prefer our bread to be white—and made from wheat.

Until the last century the harvest was a slow, back-breaking business. All day the farm-workers toiled in the fields, slowly cutting the corn with sickle or scythe. Often they went on working until darkness fell. Their wives frequently went with them to gather the corn into bundles and bind them with straw into sheaves.

Six or eight of these sheaves were then stacked against each other to form a stook. Even then the work was not finished. Each sheaf had to be turned once or twice, or 'restooked', so that all the corn dried.

A thatcher at work

Once the corn was thoroughly dry, farm-workers with two- or three-pronged pitchforks loaded it into a farm cart drawn by a horse. Then the corn was taken to a large corn barn to be threshed. This was another lengthy process. The sheaves were beaten with wooden-handled flails to separate the grain from the husk.

This large combine harvester can cut enough wheat in a day to make nearly 140 000 loaves of bread

In 1828 a Scotsman called Patrick Bell invented the first satisfactory reaper. This horse-drawn machine had huge knife blades that snipped off the corn near the ground like scissors. There were riots on some farms when this machine was used, because the workers thought it would mean that they would lose their jobs.

Later a reaper-binder was invented, which cut the corn and tied it into bundles. A few farmers still use this kind of machine. But most corn is now harvested by a combine harvester. Before the war there were only 150 in Britain; now there are over 65 000.

As the immense self-propelled machine moves across the field, it cuts the corn and threshes it automatically. A modern combine can cut and thresh as much as ten tons of grain an hour. The straw is thrown out and is ploughed back into the ground or burnt to add nutrients to the soil.

When the tank on the combine is full of grain it is discharged into a trailer or a lorry which takes it to the farm store. Because the grain is no longer dried in the sun it has to be dried artificially. This is usually done by passing hot air through the grain.

A reaper-binder. This kind of machine is now out of date, although a few are still in use

Grain crops

Two ways of storing grain—in a grain storage barn (left) and in silos (right)

There are a number of different ways of doing this. Sometimes the grain is put into large round metal storage bins, called silos, with a perforated floor through which hot air can be blown. Or the farmer can adapt one of his existing barns or buildings for grain storage. The grain is stored on the floor and hot or cold air is passed through it by means of movable ducts.

Although harvesting has become much easier, some of the fun has gone out of it, too. Many older farm-workers still remember the days when the harvest was a festival for the whole village, and men, women and children worked until late at night in the fields, returning home by the light of the harvest moon, tired but happy. And when the last of the crop had been gathered in, they all sat down together with the farmer for the harvest supper.

A 'harvest home' procession about 1820

9 The root harvest

The last big harvest on the farm is of the root crops in autumn. One of the most important of these crops is the potato. More than 90% of all the potatoes we eat are grown in this country.

Bill, the tractor driver, helped to plant the main crop of potatoes on the Essex farm in April. He was driving a special potato planter which ridges up the earth, plants the potatoes at regular intervals, and then splits the ridge to cover the seed potatoes up again. If they were not covered up by earth, they would turn green and hard and become unfit for human beings to eat.

Now it is the autumn. The potato plants have flowered and their leaves have started to wither. This means that the crop is ready to be harvested. Bill is there once again with a different tool attached to his tractor, called a potato-spinner. This machine has a wheel with a number of long, steel fingers attached to the edge, which dig into the soil and throw the potatoes to the surface.

As he starts ploughing up the ridges, a crowd of women and children follow the tractor, picking up the potatoes and putting them into baskets.

Potatoes are planted automatically by machines like this

Picking potatoes—a back-breaking job

An automatic potato picker which involves much less hard work

Although some potatoes are still harvested in this way, there are new machines which make the work easier. These potato harvesters dig up the whole crop, separate the soil from it, and deliver the potatoes into a trailer drawn alongside by a tractor. As these machines sometimes find it difficult to tell the difference between a potato and a large stone or a clump of earth, several women or men work on the machine itself to make sure that it has selected correctly.

When the potatoes have been harvested they are then stored. In the past most were stored in potato clamps. The clamp was built by levelling a piece of ground about two metres wide and the potatoes were piled up on it until they were just over a metre high. Then they were covered with a layer of straw and a layer of earth to keep out the light, which would turn them green, and the frost, which would rot them.

Although some are still stored in this way, many more are stored by simply piling them up in a special building which has insulated walls to keep out the frost.

Storing potatoes in a clamp

42 The root harvest

This machine grades potatoes. It has riddles or sieves of different sizes, through which the potatoes fall into separate sacks. The biggest, or 'ware' potatoes, are sent to market. The medium-sized are sometimes kept for replanting as new seed potatoes. The smallest, the 'chats', are fed to pigs, poultry or cattle, but when harvests are poor, they also find their way into the shops

A freshly-dug swede

On some farms there are special plants which wash or dry-brush the potatoes and put them into plastic bags ready for sale in supermarkets and shops.

Many of our potatoes come from Lincolnshire, where the dark, rich soil is particularly suitable for growing the crop. But the best seed potatoes come from Scotland and they are also grown widely in Northern Ireland and northern England.

The potato is not the only root crop grown on our farms. Two other common crops—the turnip and the swede—can be grown as food or fodder.

One of the most popular root crops today is the sugar beet. It is a triple-purpose crop. The beet is sold to factories, where the sugar is extracted; the dried pulp is then returned to the farmer for use as fodder; and the tops are fed fresh to cattle or sheep. The first modern sugar beet factory was built in Norfolk in 1912. Now there are a great number and the British farmer provides us with nearly half of our sugar.

But an increasing number of farmers now grow peas and beans instead of potatoes or other root crops. These are then sold to factories where they are used for canning or quick-freezing.

10 Intensive farming

So far we have been looking at the larger farms, ranging from the six thousand acre sheep farm to the two hundred acre mixed farm. Many farms, however, are much smaller. In fact over 40 per cent of the 228 500 farm holdings in Britain are under fifty acres.

On these smaller farms, however, the methods of farming are the same as those on the larger farms. The only difference is that less livestock is kept or fewer crops are grown.

It is only when we come to very small farms that we find a different method used. This is called intensive farming. These farms are sometimes no more than an acre or two in size and usually concentrate on growing one kind of crop or rearing one kind of animal. Thus we find special pig farms, chicken farms, and fruit farms. About 35 000 small farms are horticultural, producing nothing but vegetables, flowers, or fruit.

We shall read about these small but most important holdings in the next few chapters.

The main areas of intensive farming in Britain

Glasshouses on an intensive farm in Hertfordshire

11 The pigman

It is six o'clock on a winter's morning when John, the pigman arrives at the first of the pighouses on the intensive pig farm. This is a special *farrowing* house, where the sows (female pigs) give birth to their piglets.

A sow usually farrows twice a year. A litter may consist of 20 piglets—each weighing just over one kilo—but the average number is ten. Until it has had its second litter, the female is called a *gilt*. It will go on breeding for five or six years and may well have as many as 200 piglets altogether. Although a sow has so many piglets, it is not a very good mother.

John feeds the sows with bran mash made from grain and makes sure that the piglets are still all right. He then goes out into an open yard where the growing pigs are kept. The yard has an exit which leads into covered pens inside a pighouse.

A litter of piglets

This pen is divided by rails, because the sow might otherwise kill one or two of her litter by rolling on them. The rails are just high enough for the piglets to run under, so that they can get away from their mother. The small area of the pen outside the rails, on the left of the picture, is called the 'creep' and is warmed by infra-red lamps so that the piglets do not catch cold

Young piglets feeding from hoppers

The pigs in this section were weaned at about six or eight weeks old. Four or five different litters were put into the yard at the same time. At this stage they are allowed to eat as much dry meal as they like from a self-feeder: a round metal hopper which holds several days' food. John opens the top to see that it still contains a good supply.

Then he walks over to another pighouse where the pigs are being fattened before they are sent to market. He scatters some cubes of special fattening food on the ground. The growing pigs are split up into groups of a dozen or so and moved to the fattening accommodation when they weigh about 36 kilos.

When all the pigs have been fed he goes back for his own breakfast. Afterwards on his way back to work he looks at the male pigs, the boars who live outside in a small field. They sleep in small houses made of corrugated iron and during the day they are free to move about in the field. Sometimes, however, they are tethered by a leather harness or chain.

46 The pigman

These chains and rings are sometimes hung in piggeries to prevent the pigs getting bored and biting each others' tails

An automatic piggery

The boars all have rings in their snouts. These rings are not used for leading them, as they are with bulls, but to prevent the pigs from rooting, that is turning up the earth with their large, strong snouts. It is natural for the pig to root, since this was the way in which its ancestor, the wild boar, obtained much of its food. But rooting damages the grass and makes the field unfit for use.

John now goes back to the pighouses to clean out the sties, replacing the dirty straw with fresh. He also fills up the water troughs. The morning soon passes, and in the afternoon all the pigs are weighed. This is done once a week.

The very latest piggeries are fully automatic. The fattening pigs are fed from overhead hoppers, which drop pellets of dry food down to them at regular intervals. Or they are fed on wet mash, which is controlled by a complicated electronic switchboard. The piglets run around on a special mesh floor, which makes cleaning up their muck much easier. Modern farms become more like factories all the time!

The pigman 47

Where the cuts of bacon come from

Landrace

A pig has only to eat 3–4 kilos of dry food to put on one kilo of flesh, while a sheep has to eat 14 kilos, and a bull 20 kilos. It is therefore one of the most efficient animals for converting fodder into meat that humans can eat.

Pigs are now sold at a lighter weight than they once were, because people want lean pork or bacon with little fat. Some pigs are sold when they weigh only 40 kilos, for really lean pork which the butcher can sell at a good price. Others are fattened, mainly for bacon, until they weigh 60 to 75 kilos. The very heaviest hogs, which weigh up to 115 kilos, are used for making pork pies and sausages.

Large White

Although there are more than a dozen different breeds of pigs, the two most popular in this country are the Landrace and the Large White. As with other livestock, cross-bred animals have become increasingly popular. In 1967 a new British Saddleback breed was formed by crossing the Essex and Wessex Saddlebacks.

Wessex Saddleback

Pigs are also sometimes kept free-range on mixed farms

Pigs are one of the most useful farm animals, as they are not very particular about their food, and can eat food that otherwise would be wasted.

One food, called Tottenham pudding (or swill), is made of scraps thrown away by homes and restaurants. The pig can also eat green potatoes, which are unfit for human beings to eat.

Although the pig is one of the most common farm animals, it is also one whose habits are least understood. You have probably heard the insulting phrase 'dirty pig', but the pig is not really a dirty animal at all. It keeps its sty clean and in summer it likes to keep cool by having a shower-bath. So in hot weather some farmers spray their animals with a watering can or a hose.

Neither is the pig a particularly lazy animal. In spite of its fat it can turn at a surprising speed, as anyone knows who has tried to corner a pig.

Curiously enough, few townspeople ever think of pigs as fierce animals, although they can be savage at times. Many farm-workers and veterinary surgeons will tell you that they fear a boar, or a sow with its litter, much more than any bull.

12 The chicken farm

Like all livestock-keepers, Sam, who is a chicken farmer, starts his day by feeding his stock. As soon as it is light, he enters the deep-litter houses—long sheds in which the floors are covered with a mixture of peat and straw. This keeps so clean that it has to be cleared out only once a year. This saves Sam, who is a busy man, a great deal of work.

He gives the pullets (young birds who are in their first year of laying eggs) their morning feed of chicken mash, made of ground oats and fish meal. Then he tops up the water troughs with clean water. Finally, he puts down a little bowl of small pieces of broken flint and oyster shell for the birds to eat. This flint grit is the chickens' 'teeth'. As chickens have no teeth of their own, their food goes into the gizzard—a strong, muscular sac—where the grit grinds the food into small pieces, so that it can be digested.

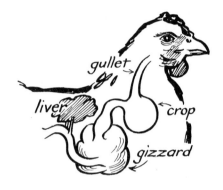

Inside a deep-litter house

Young chicks inside their shed

Next, he goes on to another shed containing the young chicks which he has just bought from a hatchery. They live under a special electric light which gives just enough heat to keep them warm. The light has to be kept on for the first five or six weeks, day and night, otherwise the baby chicks would die.

Most chickens are now reared in this way, though at one time they were all reared by the mother hen, who allowed the chickens to nestle under her warm body and soft feathers. Her body temperature is about 42°C.

Sam gives the chicks their special chick feed and then makes sure that none of them are huddling together. This is very important, for if the baby chicks huddle together they suffocate and die.

Then he goes out into the fields, where the younger pullets are kept in special chicken sheds. He lets them out and scatters grain on the ground, making sure that he spreads it widely enough, so that they all have a fair chance to eat.

A hatchery. All the boxes on the shelves in the background are full of chicks

It is not only healthier for growing birds to be out in the open on free-range, but they can also pick up some of their own food from the ground in the form of insects and grass. There is nothing a chicken likes better to eat than a fat, juicy worm!

These younger pullets are only two months old, so they will not be laying eggs for at least another three months. Sam will know when they have reached the point of lay because the combs on top of their heads, which are now pink, will start to redden. Then he will move them into the laying houses. The feeding is over now and Sam can go home for his own breakfast. (As you have seen, the animals always come first on all livestock farms.)

The birds have to be fed again in the early afternoon and at dusk the outside birds have to be shut up for the night. If this were not done, Sam would soon lose his birds to the foxes.

Spring is one of the busiest times of the year for Sam, as most of the young birds are being reared at this time.

Free-range hens

52 The chicken farm

Light Sussex

Autumn is another busy season. This is when some of the birds go into moult, that is, they shed their feathers which are replaced by new ones. When this happens, the pullet becomes known as a hen. In this moulting season the bird stops laying eggs. Most hens do not lay so many eggs the second year, so that Sam has to decide which ones to keep. He sends most of the birds to the butcher, who sells them as boiling hens.

There are many different breeds of chickens. Some of the best known are the Rhode Island Red which lays tinted or brown eggs; the Leghorn, a prolific layer of white eggs; and the Plymouth, White and Buff Rocks which are good *table birds* for eating.

No single breed, however, produces both cockerels which are good for eating and hens which lay a large number of eggs. Most farmers therefore keep cross-bred birds.

The Leghorn is often crossed with a Rhode Island Red for egg production, and the North Holland Blue with a Light Sussex for table poultry.

Most laying hens in Britain are now kept separately in batteries of small wire cages. The floors are made of sloping wire netting or mesh, so that the eggs roll down into a rack in front of the cage. The birds are usually fed and watered automatically.

Rhode Island Red

Buff Rock

North Holland Blue

Some farmers are not interested in producing eggs, but 'broilers', or table chickens, which will be sold frozen in a plastic bag in a supermarket.

The number of these 'factory farms', as they are sometimes called, has increased enormously in the last few years, to satisfy the increasing demand for roasting chickens. Chicken is no longer a luxury food for Christmas, as it once was. We now eat nearly 15 kilos each of chicken meat in a year—five times as much as we did thirty-five years ago.

The chickens are kept in special houses, which hold 5000 to 15 000. They are put into the houses as day-old chicks. They are killed when they are nine or ten weeks old and weigh around $1\frac{1}{2}$ kilos. Each house has a huge outside food bin into which the food is delivered by tanker. The birds are fed and watered automatically. As the houses have no windows, the birds live in artificial light, which is also automatically controlled. Some farmers keep the lights on the whole twenty-four hours.

At nine or ten weeks they are taken to a factory where they are plucked and cleaned by machine and are then put into plastic bags and frozen.

On some poultry farms eggs are collected on a trolley at regular intervals. In more modern battery houses they are transported on a moving belt to the egg room where they are graded and packed

The Red Jungle Fowl of India—the ancestor of all our domestic hens

13 Fruit and vegetables

Many of the fruits and vegetables that we eat are grown on British farms.

Many of the vegetables come from mixed farms, where whole fields of potatoes, beans, and peas are grown. But we also obtain a large quantity from smaller intensive farms, called horticultural holdings. Through experience it has been found that some vegetables grow best in certain parts of the country where both the climate and the soil are favourable. Thus, many of our early potatoes and winter cauliflowers come from Cornwall; carrots from Norfolk; celery and onions from the Fen district; brussel sprouts and onions from Bedfordshire.

Some horticultural holdings grow mainly the three chief salad crops: tomatoes, cucumbers, and lettuce.

Two of the biggest tomato-growing areas are the Lea Valley near London and the district around Worthing in Sussex. Like the potato, the tomato came originally from South America. The cucumber, however, was brought into England from India.

Brussel sprouts grow on long tall stalks

A bumper crop of tomatoes in a glasshouse

A horticultural holding, growing flowers and fruits

On most holdings many of the crops are grown under glass. Horticultural glasshouses are of two main types: the low-roofed *vinery* used for lettuce, flowers, and plants; and the higher-roofed *aeroplane*, in which tomatoes and cucumbers are grown. They are heated by hot water pipes or by electricity.

Lettuces and seedlings are often raised under cold frames, so-called because they are not heated. Built of metal, wood, or brick, they are covered by a glass frame called a *light*. The glass provides just that little extra amount of protection to allow seeds to germinate early.

Although our vegetables come from both mixed farms and horticultural holdings, much of our fruit comes from special fruit farms in Kent, Norfolk, Cambridgeshire, Essex, Suffolk, and Sussex.

Fruit and vegetables

Spraying fruit trees in an orchard. Notice that the tractor driver is wearing a mask to stop him inhaling the poisonous chemical fumes

Pruning old apple trees

Kent is, perhaps, the best known county for its fruit, particularly apples, pears, and cherries. On an apple farm in Kent, the first thing you would notice is that the trees are set out in neat rows, instead of being dotted about all over the place. There is just enough space between the trees so that when each is fully grown the leaves will not touch.

There are also some beehives between the rows of trees. Bees perform very useful work on the fruit farm. They carry the pollen from one tree to another, *pollinating* the blossom, so that the fruit will grow.

Early summer is the most worrying time for the fruit farmer, because there are sometimes late frosts. By then the delicate blossom has appeared and one severe frost could destroy the whole crop. Once the first fruitlets have appeared, many of the smaller fruitlets are cut off by the farmer with small shears. This is done because a tree can grow only a limited amount of good fruit. The farmer would rather have a reasonable crop of fine fruit than a large crop of poor fruit that no one wanted to buy. The trees are then sprayed with chemicals to prevent insects from damaging the fruit.

Once the fruit starts to ripen, the surrounding leaves are carefully trimmed away so that the sun can ripen the whole fruit.

The crop is harvested by hand. Students are often employed in their holidays. But because it is so difficult and so expensive to get people to do this work, many farmers have given up growing such crops as gooseberries and plums. More and more farmers let customers come and pick the fruit themselves.

Apple pickers at work

Hop picking the modern way. The machine strips the hops off the 'vines'

One crop widely grown in Kent is hops, which are used in brewing beer. The plants are trained to grow up wires or strings, and grow as fast as a beanstalk, climbing 4·5 metres or more in only seven weeks.

Near the hop-garden you will often see an oast-house. There are many of these picturesque buildings with their cone-shaped roofs in Kent. They were used for drying hops over a charcoal fire, while the fumes drifted gently out of the tall chimney. But the modern oasts are not nearly so pretty. They are small, squat buildings with flat roofs, containing kilns heated by oil burners.

Before we leave the intensive farms, one more must be mentioned, as it is the only farm you will ever find underground. At one time all mushrooms used to grow wild in fields where horses grazed, because their manure suited this crop. As most horses have now been replaced by tractors, mushrooms have to be grown artificially. And because they grow well in the dark, some mushroom farms have been started underground in disused railway tunnels.

An underground mushroom farm

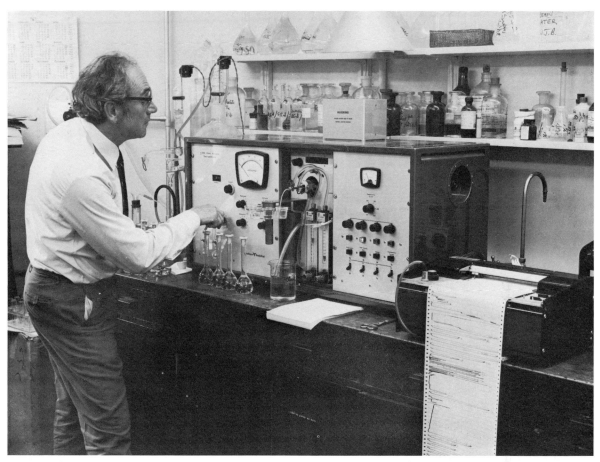

This scientist is analysing soil samples. Other scientists do experiments to breed new plants, increase the weight of cattle, or to discover the best plants for each kind of soil—all to make farming a more efficient and economic business

14 New plants for old

All our vegetables and flowers once grew as weeds. The carrot, for example, is descended from a poisonous plant with coarse roots which still grows wild.

Improvements have been made by the careful selection of the seeds from the best plants, often over the course of centuries. At one time man had to wait for nature to make improvements and many advances in farming came by chance.

Once, when a Suffolk farm-worker was returning home after threshing, he found some seeds of barley in his boots. Noticing that they were of a particularly good shape, he planted them in his garden. The crop was excellent. He took the best seeds, grew more plants and saved the best seeds again. His patient work eventually produced a fine, new strain of barley, called Chevallier.

Blighted potato leaves

Plants can be crossed, or bred together artificially, to produce a new hybrid or mixed form. In this way, the best features of two completely different plants can be combined into one.

The work, however, needs great care. The male parts of the flower, the *stamens*, must be carefully removed, and the female organ, the *pistil*, pollinated by the stamens of the other plant. In this way, scientists have produced wheat which resists some diseases, and loganberries without thorns.

This is only one of the many ways in which scientists are helping the farmer. Another is in the control of pests, such as beetles, slugs, snails, and bugs. For many years farmers were puzzled why corn, which was sown after a ley (see page 26) had been ploughed in the autumn, was often attacked by leather-jackets, the grubs of the daddy long-legs. Scientists studied the life-history of the pest. They found that it always laid its eggs in the grass in August and they told farmers that, if they ploughed up their grass before then, they would have no more trouble with leather-jackets. And they were right.

Wireworm

New plants for old   61

Woodlouse

Codling moth larva emerging from an apple

Snail

Slug

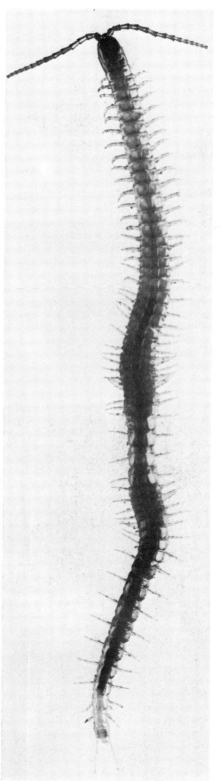
Centipede

Spraying a crop of beans by helicopter

Scientists have invented insecticides to kill other pests. The farmer sprays his crops with various insecticides, which kill the pests, yet do not harm the growing crops. This is sometimes done from the air by a helicopter.

Other chemicals are used to kill weeds. Some weed-killers are selective. They kill only the weeds and leave the corn or grass unharmed.

The scientist also supplies the farmers with plant foods in the form of artificial fertilisers, made from chemicals. Computers are now used to help breeders decide which pigs will produce the greatest amount of meat most cheaply.

The first farming research station was opened by Sir John Lawes in his home, Rothamsted, Hertfordshire, in 1843. Today, the Rothamsted Experimental Station is only one of more than forty Government-supported research stations where scientists carry out experiments to find new ways to help the British farmer.

15 Helping the farmer

If you visit a farm, there are certain things you can do to help the farmer:

1 *Close gates* after you.
2 *Be quiet* if you are allowed to go into the milking shed to see the cows milked. In fact, you should be quiet whenever you are near animals. Most animals are easily frightened.
3 *Keep to the footpaths* when you cross fields.
4 *Keep away from machinery*.

Finally, here are a few 'don'ts':

1 *Don't leave litter*. This is not only because it looks untidy. Pieces of glass or old tins left on the ground may easily be trodden on by animals and hurt them.
2 *Don't climb over fences* or barbed wire. Always use the gate or stile.
3 *Don't throw things into ponds*. Animals very often drink the water from ponds.
4 *Don't light fires*, unless you are given permission.

A gate left carelessly ajar

Some other books about farming

Dark, Judith C, *Farming in Britain*, Ginn, 1972
Toulson, Shirley, *Discovering Farm Museums and Farm Parks*, Shire Publications, 1977
Haines, George Henry, *How We Find Out in Agriculture*, John Baker, 1971
Huggett, Frank, *A Day in the Life of a Victorian Farm Worker*, Allen and Unwin, 1972
Sauvain, Philip, *On a Farm*, Macmillan Education, Environment Books, 1974

A herd of bullocks, bred from a Hereford bull and Friesian cows

Index

arable farming 5, 7, 8, 9, 25, 26
apples 8, 56, 57

barley 9, 18, 26, 28, 36, 59
beans 27, 42, 54
beef farming 9, 17–19
bees 56
beer 8, 36, 58
Bell, Patrick 38
brussels sprouts 54

cabbage 26
carrots 54, 59
cattle 6, 7, 10, 16, 18, 25, 26, 27, 28, 33
   Aberdeen-Angus 19
   bullocks 6, 9, 17, 18
   calves 17, 22, 23, 25
   Charolais 19
   Friesian 18, 20, 23
   Galloway 10, 19
   Guernsey 23
   Hereford 19
   Jersey 23
   North Devon 6
   Simmental 19
cauliflowers 54
chickens, *see poultry*
Coke, Thomas 28
combine harvester 38
Common Market 25
corn 29, 30, 31, 36, 37, 60
cucumbers 8, 54, 55
cultivator 30

dairy farming 5, 6, 7, 9, 20–24, 25, 26
disease 12, 24, 26, 60–62

eggs 52, 53

factory farms 53
fertilisers 28, 62
flowers 43, 55, 59
foot-and-mouth disease 24
fruit farming 8, 43, 54, 55, 56–57

glasshouses 8, 55
grain crops 9, 25, 30, 36–39
grass 7, 10, 17, 18, 21, 26, 33–35, 62

harrow 7, 30–31
harvest 25, 33–35, 37–39, 40–42, 57
hay 11, 17, 21, 27, 28, 32, 33–35
hill farming 6, 9, 11–15
hops 8, 58
horses 6, 7, 29, 32, 36, 37

insecticide 56, 62
intensive farming 8, 9, 43, 44–58

kale 25, 26

lettuce 54, 55

machinery 3, 7, 9, 12, 21, 22, 29–32, 33, 34, 38, 40, 41, 62
milk 5, 6, 17, 20–23, 25
mixed farming 5, 6, 7, 8, 9, 24, 25–28, 29
mushrooms 58

Norfolk rotation 28
Northern Ireland 17, 42

oast-houses 8, 58
oats 26, 27, 28, 36
onions 54

pasteurising 22
peas 27, 42, 54
pests 12, 60–62
pigs 5, 10, 25, 26, 42, 43, 44–48, 62
   Landrace 47
   Large White 47
   Saddlebacks 47
ploughing 19, 29–32
ponies, *see horses*
potatoes 7, 9, 25, 26, 40–44, 48, 54, 60
poultry 16, 25, 26, 43, 49–53
   Buff Rock 52

Leghorn 52
Light Sussex 52
North Holland Blue 52
Rhode Island Red 52

reaper-binder 38
Red Jungle Fowl 53
research 59–62
root crops 9, 21, 28, 40–42
rotation of crops 26–28
rye 27, 37

Scotland 10, 13, 15, 16, 17, 19, 24, 36, 42
seed drill 31
sheep 5, 6, 7, 9, 10, 11–16, 17, 18, 25, 27, 47
   Border Leicester 16
   Cheviot 15
   lambs 5, 6, 11, 12, 13, 16
   Lincolnshire 15
   Oxford Down 16
   Scottish Blackface 15
   Shetland 16
   Southdown 7, 15
   Wiltshire Horn 16
sheepdogs 11, 14–15
shepherd 5, 11–15, 16
silage 35
silos 35, 39
sugar beet 9, 42
swedes 42

thatch 37
tomatoes 8, 54
tools 12, 30, 37, 40, 56
Townshend, 'Turnip' 28
tractors 3, 29–32, 34, 40
turnips 11, 25, 27, 42

vegetables 9, 25, 43, 54–55, 59

Wales 13, 17, 19
wheat 9, 25, 26, 27, 28, 36, 37, 43, 60
wool 6, 12, 15, 16

James Simpson and Chloroform

To M.V.A.

I am crucified with Christ
Nevertheless I live
    Galatians 2 : 20

Pioneers of Science and Discovery

James Simpson
and Chloroform

R. S. Atkinson, M.A., M.B., B.CHIR.,
F.F.A.R.C.S.

Consultant Anaesthetist, Southend-on-Sea Group of Hospitals

PRIORY PRESS LIMITED

Other Books in This Series

Carl Benz and the Motor Car Douglas Nye
George Eastman and the Early Photographers Brian Coe
Alexander Fleming and Penicillin W Howard Hughes
Marie Stopes and Birth Control H V Stopes-Roe
Joseph Lister and Antisepsis William Merrington
Edward Jenner and Vaccination A J Harding Rains
James Simpson and Chloroform R S Atkinson
Louis Pasteur and Microbiology H I Winner
Richard Arkwright and Cotton Spinning Richard L Hills
Alfred Nobel, Inventor of Explosives Trevor I Williams
William Harvey and Blood Circulation Eric Neil
Isaac Newton and Gravity P M Rattansi
Rudolph Diesel and the Power Unit John F Moon
Michael Faraday and Electricity Brian Bowers

Frontispiece Portrait of Sir James Young Simpson, M.D.

SBN 85078 119 1
Copyright © by R. S. Atkinson
First published in 1973 by
Priory Press Ltd, 101 Grays Inn Road, London, WC1
Set in 'Monophoto' Baskerville and printed offset litho by
Page Bros (Norwich) Ltd, Norwich

Contents

	Preface	8
1	Childhood and Medical Studies	11
2	The Successful Medical Man	21
3	The Discovery and Development of Anaesthesia	29
4	Chloroform	41
5	Edinburgh and London	63
6	Home and Abroad	79
7	Simpson's Later Years	85
8	Edinburgh Today	89
	Date Chart	90
	Further Reading	91
	Glossary	92
	Index	93
	Appendices	95–6
	A. *Part of the Family Tree*	
	B. *Chemical Formulae*	
	C. *Professors at the University of Edinburgh*	

List of Illustrations

Portrait of Simpson	*Frontispiece*
Anaesthetic equipment today	10
The Royal Infirmary, Edinburgh	12–13
Roadsign, Bathgate	14
Amputation before anaesthesia	17
Eighteenth century Hospital	20
The House of Commons, 1832	23
Sir Humphrey Davy	28
Newgate prison	28
Naval discipline	30
Alcoholic fumes as an anaesthetic	30
William Morton administering ether	32
The first public administration of ether	33
An operation without anaesthesia	34
Thomas Wakley	35
A paddle steamer	36
An experiment with ether	37
Amputation in the Tudor period	38
Simpson experimenting on himself with chloroform	40
52 Queen Street, Edinburgh	41
The Discovery Room, 52 Queen Street	43
The Discovery of chloroform	44
18th century Apothecaries' advertisement	48
Letter by Simpson	51
Florence Nightingale	52
The first death under chloroform	53
Schimmelbusch mask	55
Simpson's pill box	57
Simpson's stethoscopes	58
Simpson's cross	62
Title page of Davy's book	64
Davy's "gas machine"	65
Modern face masks	66
John Snow, M.D.	69
A page from Snow's diary	70
Queen Victoria	71
Another page from Snow's diary	75
Punch's view of chloroform	77
A modern anaesthetist	78
A lung operation	83

Sir Joseph Lister	84
Simpson's grave	87
Simpson's statue in Edinburgh	88

Picture Credits

The author and publishers wish to thank all those who have given permission for copyright illustrations to appear on the following pages: Mary Evans, 17; Kodak Museum, 71; Keystone, 10, 78; Mansell Collection, 2; National Portrait Gallery, 52; Pictorial Press, 83; Ronan Picture Library, 33, 37, 40, 53, 77; Mr. R. Hamilton, 43, 57, 62; the Royal College of Physicians, London, 70, 75. All other photographs are the property of the author or of the Wayland Picture Library.

Preface

The discovery of anaesthesia was made in the New World, but it was in Britain that it was developed. Men like John Snow in London and James Simpson in Edinburgh studied and wrote about it, advocating proper methods of administration of ether and chloroform, and ensuring that, in this country at least, the specialty was developed by the doctors. In Britain, right from the start, anaesthetics were administered by physicians who were also anaesthetists, in contrast to the practice elsewhere of delegating the task to nurses, porters or others.

The story of Simpson's major contribution, the discovery of chloroform anaesthesia, is a fascinating one. It has been told in a number of biographies. The purpose of the present volume is to set out the main events of the story in a simple way.

The author acknowledges the material obtained from earlier and more complete biographies listed in the Bibliography. I would like to thank the many people who have helped me with their advice. I have received considerable help and encouragement from my colleague, Dr J. Alfred Lee. I am grateful to Mr J. K. Wilson, MPS, Mr Nicholas Herdman, MPS, and Dr W. T. Simpson, who were all associated with Duncan, Flockhart & Co. Ltd., and who have given me information regarding the early history of the company, and to Mr I. B. Smith, M.P.S., of B.D.H. Pharmaceuticals for permission to use the History of Duncan, Flockhart and Co. and to reproduce Simpson's letter. Mr R. Hamilton, Leader of the Committee on Moral Welfare of the Church of Scotland, was also kind enough to show me round Simpson House and he provided a number of photographs. I wish also to thank Mr John Wood for his photographic help, and Miss I. Gaunt and Mrs J. Bramley for their able secretarial assistance. I should also like to thank my wife for her forebearance.

The importance of anaesthesia can be seen in this picture of an accident ward today.

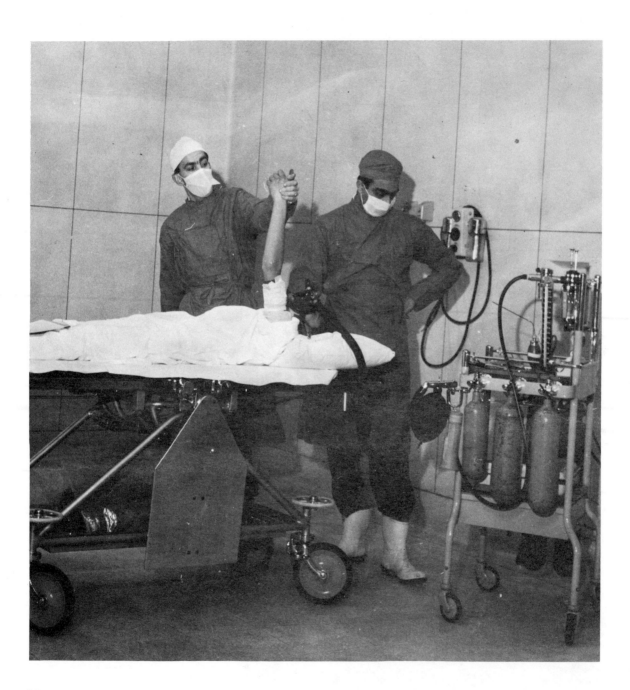

1 *Childhood and Medical Studies*

James Simpson was born on 7th June, 1811, in the town of Bathgate, in what is now West Lothian. He was the seventh son and eighth child of his parents, David and Mary Simpson. The entry in the Register of Births and Baptisms for the Parish of Bathgate states that he was baptized on 30th June and given the name James. The name Young was added later, perhaps because on account of his youthful appearance he came to be known as "Young Simpson" by his Edinburgh friends.

His grandfather, Alexander Simpson, was a farmer and farrier in the district. Alexander's wife, Isabella, had the maiden name, Grindlay, and it was her great niece, Jessie Grindlay, who was later to marry James. Alexander Simpson was noted for his skill in the treatment of animal diseases. When orthodox methods failed, country superstitions demanded strange measures akin to witchcraft. Thus it is related that when cattle plague was prevalent in the late eighteenth century, Alexander, with the help of his youngest son David, buried a cow alive as the only cure against the evil eye. The latter was to remember for the rest of his life how the ground heaved after the cow had been covered with earth.

Alexander and Isabella had five sons. David worked as a journeyman in London, Glasgow and Leith, before returning to his homeland. He set up a whisky distillery and in 1792 married Mary Jarvie, the daughter of a neighbouring farmer. Mary was descended from one of the Huguenot families who had fled France after the St Bartholomew's Eve massacre of Protestants. For a time the family prospered and a succession of children were born. But new excise laws made whisky production less profitable and David turned to the making of bread, setting up in business in a house in Main Street, Bathgate, in 1810.

Main Street was then on the main traffic route between Glasgow and Edinburgh. The bakery, however, was not at first a success. The business got into debt and Mary became pregnant again. The baby was born on 7th June, 1811. Dr Dawson recorded the event:

> June 7. Simpson, David, baker, Bathgate.
> Wife, Mary Jarvie. AE 40. 8th child, son.
> Natus 8 o'clock. *Uti Veniebam natus.*
> Paid 10s. 6d.

The baby James was the eighth child, but the seventh son. The family fortunes were at their lowest ebb when he was born. That day the doctor's fee was 10s. 6d., but the shop takings were only 8s. 3d. However, there were strong beliefs in the luck and good fortune brought to a seventh son. Be that as it may, the family fortunes began to recover from this time. This was largely because Mary Simpson took an active part in the management of the bakery. Soon the family was able to move to a house on the other side of the street.

The youngest child of the family received a great deal of affection from his parents and from his older brothers and sister. He is said to have been a small, thickset and sturdy child with curls surmounting a large head. He was encouraged to read and learn, starting school at the age of four years. Education

The New Royal Infirmary, Edinburgh. This replaced the hospital in which Simpson's clinical training took place.

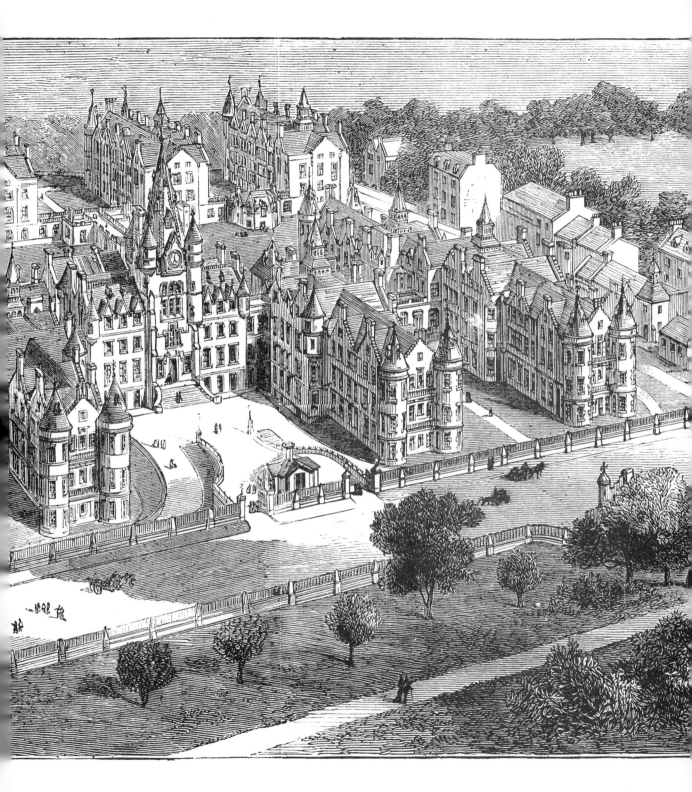

This road sign is seen by visitors as they enter Bathgate.

was something of a privilege in those days and it was often a family's ambition to send one child to university. In the Simpson family James seemed to be earmarked for this distinction. He did well at his school work and took an interest in the countryside. His mother taught him the simple Christian beliefs, and his great uncle, George Jarvie, who kept one of the local inns, planted the seeds of his lifelong interest in archaeology.

When James was nine years old his mother died. His elder sister, Mary, took over the management of the household and continued to lavish love and encouragement on her young brother. About this time his great uncle George took him on his first visit to Edinburgh. The young lad must have been excited by his first view of the city with the castle overlooking the roofs and spires of the town buildings. They spent time in the Greyfriars churchyard and copied inscriptions from the tombstones.

The house in which James Simpson was born has subsequently been used as a mission hall. Both this and the houses opposite have now been demolished, the mission hall as recently as 1969. Today, signposts on all roads leading into Bathgate proclaim the town as the birthplace of Sir James Young Simpson, Discoverer of Chloroform. Sadly, there is no statue, memorial or plaque in Main Street. The street itself is a relative backwater, new roads now carrying the traffic through another part of the town. The house has been demolished and the site cleared, serving now only for the parking of motor cars.

James Simpson began his studies at the university in Edinburgh at the age of fourteen. He entered the arts course and lodged at 1 Adam Street, together with one John Reid. John Reid, his senior by two years and a former companion at Bathgate, no doubt set him a good example. He was now a medical student. Also in the house was a Mr

MacArthur, once a junior master at Bathgate school, and now also engaged in the study of medicine. MacArthur was an early influence on Simpson, encouraging him to persevere with the course. MacArthur believed that time spent sleeping was time wasted. He is said to have related that he himself could manage with four hours sleep a night and John Reid with six, but that he had not yet broken in the young James Simpson. MacArthur did not become well known, but his two protégés testify to the value of his influence during a formative period of their lives. Reid was to become Professor of Physiology at St Andrew's University; Simpson was to achieve fame in more than one sphere and was never to forget the lesson that time was a valuable commodity, not to be wasted.

The university scene at this time was frequently turbulent. There were arguments between rival professors, but nothing was more notorious than the scandalous affair of Burke and Hare, which involved the great anatomy teacher, Robert Knox. It was towards the end of 1828, when Simpson was just about to begin his medical studies, that the storm broke.

Burke and Hare were body snatchers, who broke open graves soon after interment and removed the corpses for sale as anatomy material. The science of body snatching was developed to a remarkable degree, special tools being employed for silent entry to the grave, while respectable citizens erected fences or cages around the tombs in attempts to prevent such vandalism. Matters took an even more serious turn when Burke and Hare took to murder in order to provide enough bodies for the anatomy rooms. When they were discovered Hare turned King's Evidence, helping to convict Burke who was duly hanged. It came to light that the majority of the bodies had been disposed of to Knox and a fierce outcry against him spread through the city.

This patient looks amazingly placid during an amputation performed before anaesthesia had been invented. In reality, the pain of this and other operations was terrible.

There were times when the crowds besieged the classrooms shouting against Knox. Simpson did not come into the class until some time later, about 1830–31. John Reid became one of Knox's demonstrators in 1833 and Knox visited Reid's family at Bathgate. However, Knox never recovered from the Burke and Hare scandal. He finally became demoralized and left Edinburgh for London where he died some years later almost destitute.

Edinburgh at this time was one of the leading medical schools in the world and a centre of intellectual activity. Simpson was indeed lucky to be educated in this stimulating environment. The affairs of the university were controlled by the Town Council and there was little disciplinary control of the students. The system of leaving a student much to his own devices probably had an excellent effect on men of Simpson's calibre and energy. He was also able to sit at the feet of some of the most renowned surgeons and physicians of the time. These included Robert Liston, the surgeon who later moved to University College Hospital in London and who was to perform the first major surgical operation under ether anaesthesia in Britain. The Professor of Midwifery was James Hamilton. Here again credit is due to Edinburgh, for the teaching of this subject goes back to the year 1726. The Edinburgh Chair of Midwifery was indisputably the first in Britain, and probably in the world. Nevertheless it remained a Cinderella of the medical faculties for a long time. Thomas Young, professor in 1756, was the first to teach the subject to medical students by lectures and clinical teaching. James Hamilton was the first professor to obtain recognition of midwifery as a subject necessary for obtaining the university medical degree. It was to be Simpson who raised it to a position of parity with the other faculties, leaving it a science whose professors were some of the most distinguished among the medical fraternity.

Simpson was a diligent, though not exceptional, medical student. It is said that of the various lectures he attended, those on obstetrics were least interesting. He often fell asleep during Hamilton's lectures. His first contact with operative surgery was something of a shock. When Liston carried out a breast amputation without any form of anaesthesia (as none was yet invented), he rushed away from the hospital to enrol as a law student. Simpson later recalled this moment and asked himself, "Can nothing be done to make operations less painful?" Fortunately for posterity he returned to his medical studies. In 1830 he was due to present himself for the final examination of the College of Surgeons. In January of that year he was recalled to Bathgate where his father had been taken seriously ill. He returned to care for him in his final illness, and after his death returned to Edinburgh for the examination which he passed with credit. In April 1830, at the age of eighteen, he was a qualified medical practitioner.

2 *The Successful Medical Man*

Although a member of the Royal College of Surgeons of Edinburgh at the age of eighteen, he had to wait until he was twenty-one to proceed to the MD. Conscious of the need to relieve his family of the burden of supporting him, he applied for a post as parish surgeon in the village of Inverkip on the Clyde. To his chagrin he was not appointed. Instead, with the help of his elder brother Alexander ("Sandy") he returned to Edinburgh for further study, boarding with his brother, David, who had become a baker at Stockbridge. He became an assistant to a Dr Gairdner in dispensary practice but his main object was to obtain the MD qualification which he did in 1832. Simpson was among the last candidates to be examined in the Latin language; this became optional the next year. The Professor of Pathology, Dr John Thomson, was impressed with the young Simpson and offered him the coveted post as assistant with a salary of £50 a year.

James Simpson took on his new post with enthusiasm and became almost indispensable to his chief. He worked hard in the museum and took up the new instrument, the microscope. Thomson was responsible for Simpson's change to the subject of midwifery. The reason for this is not clear, but Thomson had been dubbed "the old Chairmaker" by Knox, and it may be that he appreciated the potential of the young man. At any rate, Simpson attended the midwifery lectures during the winter session of 1833–34, this time with new zeal.

Despite the interest in midwifery, Simpson had to

Although this picture shows an eighteenth century hospital, the wards in which Simpson worked as a student would have looked very like this.

devote most of his time to his duties in the pathological department. Pathology was a new science: the Chair had been instituted only in 1831 and Thomson was the first professor. Thomson, however, was a man of no small stature. He had previously held Chairs of General and Military Surgery and had visited the field of Waterloo after the battle. He was also a philanthropist who set up the New Town Dispensary in 1815 in the face of considerable opposition. Such was the man who influenced Simpson's career and sparked off his interest in the hardships of the poor.

The Royal Medical Society was the oldest society in the university, having been established in 1737. Simpson was elected a member of this and of the Royal Physical Society. He was widely read and a frequent speaker, asking questions and storing details in his brain. He and the intellectual Edward Forbes became associated. They were interested in all current topics, supporting the Reform Bill which was before Parliament in London, taking up the anti-slavery cause, and joining those who mourned Sir Walter Scott to consider a monument in his memory. Another young doctor whom Simpson came to know was Dr (later Sir Douglas) MacLagan. In 1835, financed by his older brothers, James Simpson embarked on a continental tour with MacLagan. The two Edinburgh doctors visited museums and hospitals in London, heard Peel speak in the House of Commons, and then set off by coach to Southampton on the way to Paris. The French doctors demonstrated cases and then they returned via Liège, Birmingham, Liverpool and Glasgow.

The visit to Liverpool has a special significance. A few years previously James's brother John had written to a relation in Liverpool, one Mr Walter Grindlay, to see whether a temporary post as ship's surgeon could be found. At the time this came to

Simpson visited the House of Commons during his journey to the Continent.

nothing, but James was now to visit these relatives. Simpson entered in his diary, "July 6th. At 7 I set off to drink tea and spend the evening with Mr Grindlay and his family . . . one of the Misses Grindlay has a resemblance to Mary. Much more like that of a sister than a second or third cousin." During this brief visit friendships must have been cemented. Mr Grindlay helped Simpson with loans of money at various times. Jessie Grindlay was to become his wife.

James now felt that the more strictly student part of his life was at an end, and that he had matured sufficiently to enter into competition with his professional brethren. In November 1835 recognition came when he was elected one of the annual presidents of the Royal Medical Society. Previous presidents had risen to fame, and James was anxious not to let the opportunity slip by. He took great pains with the inaugural address. He took as his subject "The Diseases of the Placenta" and spent much time and energy amassing references and obscure pieces of information from old books. He was rewarded by the reception his paper got. It was published in the *Edinburgh Medical Journal* in 1835, and shortly afterwards translations appeared in France, Germany and Italy. He was made a corresponding member of the Ghent Medical Society. From now on, Simpson was recognized as an authority on the subject.

We have been favoured by a description of the man at about this time. A visitor to the Society wrote:

> The chair was occupied by a young man whose appearance was striking and peculiar. As we entered the room his head was bent down, and little was seen but a mass of long tangled hair, partially concealing what appeared to be a head of very large size. He raised his head, and his countenance at once impressed us. A pale, rather flattish face, massive bent brows, from under which shone eyes now piercing as it were to your inmost soul, now melting into almost feminine tenderness;

a coarsish nose with dilated nostrils, finely chiselled mouth which seemed the most expressive feature of the face...

At a time when physicians and surgeons cultivated peculiarities of appearance and behaviour, Simpson had a presence which did not require artificial aid.

Simpson's clinical practice was also growing, though it was scattered over the city and involved considerable walking. He moved from his brother's shop to take rooms in Teviot Row, later moving to 2 Deanhaugh Street, before acquiring a residence of his own by renting a house, 1 Dean Terrace. His professional work kept him fully occupied. Nights were disturbed by the calls of his obstetric patients. He often rose early, sometimes even at three o'clock, to compose his medical papers which were being published in the journals in increasing numbers. In 1836 he became house surgeon at the lying-in hospital for twelve months to increase his clinical experience. When Thomson became ill he assumed the post of Deputy Professor of Pathology. His sights, however, were on the obstetric horizon. This was to be his career in life, though its start was delayed by his responsibilities to the Department of Pathology. It was not until 1838 that he became a lecturer on midwifery.

James kept up his friendship with John Reid. Both were elected Fellows of the Royal College of Physicians of Edinburgh. The two friends attended parties and dances and together they made a journey south, visiting Liverpool, and returning by boat. James was writing letters to Jessie Grindlay with increasing frequency. When Jessie travelled north, James wrote to his brother, Sandy: "Tell Mary that Miss Grindlay goes to Boness next week. If she has a spare bed, I should like greately to offer her a night's entertainment in Bathgate. She is very anxious to know Mary, and is really an excellent creature...." But a subsequent letter shows that the occasion was missed: "Tell Mary that Miss Grindlay is at Boness. She went yesterday; but I persuaded her past her Bathgate trip this time." James, however, was soon able to visit Liverpool again. He received an invitation to visit Dublin in the autumn of 1839. The doctors there entertained him well, and he was able to visit the Grindlays on his way home.

Now Professor Hamilton resigned the Midwifery Chair. Simpson applied in a letter from Dean Terrace dated 15th November 1839. His hopes could not be considered great. He had been told that his youth and celibacy would count against him. Dr Lee and Dr Kennedy of Dublin were serious rivals. However, James resolved to make every effort to obtain the coveted Chair.

The appointment was in the hands of the Town Council, not of the university. There were thirty-three members of the Council. It was the usual practice in those days to canvass all the members. The professors of the other medical faculties also held strong views regarding appointments and tried to exert their influence in favour of their own choice. Simpson's advantage lay in his achievements. His name was known to all readers of the medical journals both at home and abroad. His biggest disadvantage was that he was unmarried. Nobody of consequence would be so indelicate as to allow his wife to consult an obstetrician who was a bachelor. James remedied this defect. On 26th December 1839, he married Jessie Grindlay, having slipped away to Liverpool. Mr Grindlay had given his consent despite the fact that his intended son-in-law was in debt to his brother Sandy for a sum in excess of £300. There was no honeymoon. The election campaign was at a critical point and James could not be away from Edinburgh. Jessie had to help him catalogue the contents of his museum, it being essential for a candidate to have his own collection of specimens and teaching material.

The election candidates gradually withdrew, until only two were left in the field. Simpson's rival was Dr Evory Kennedy of Dublin, a strong opponent. Simpson spent £500 in the campaign, a large sum even taking into account the cost of printing and postage at that time. However, testimonials had to be much more numerous and elaborate than is expected today. Kennedy's made a volume of 150 octavo pages!

The day of election fell on Tuesday, 4th February 1840. All thirty-three members of the Council were present. The Provost proposed Kennedy and Baillie Ramsay proposed Simpson. Simpson won by seventeen votes to sixteen. That evening James wrote to Liverpool to convey the glad tidings to his father-in-law. James and Jessie commenced their delayed honeymoon the next day.

Sir Humphrey Davy, the discoverer of "laughing gas" (nitrous oxide) which was used as an anaesthetic before the discovery of chloroform (*left*).

The conditions in Newgate prison show something of the callous attitude of the period (*below*).

3 The Discovery and Development of Anaesthesia

On 16th October 1846 an event took place in the Massachusetts General Hospital, Boston, USA, which was to have the most profound significance for the world. On that day William Morton administered ether to a patient, Gilbert Abbott. While the patient slept, the surgeon ligated a vascular tumour of the neck without causing pain. Anaesthesia was born. He announced, "Gentlemen, this is no humbug," and Henry J. Bigelow, who was a spectator, made the prophetic remark, "I have seen something today which will go round the world."

The discovery of anaesthesia was not, however, the flash of inspiration of one particular individual. Humphrey Davy working on nitrous oxide around 1800 had said, "As nitrous oxide in its extensive operation appears capable of destroying physical pain, it may probably be used with advantage during surgical operations in which no great effusion of blood takes place." How was it that such words were not acted upon by the surgeons of the day? It must be remembered that Britain was engaged in the worst war of the country's history. Suffering in various forms was regarded as a necessary part of life. Soldiers and sailors were flogged. Prison conditions were barbaric. Anaesthesia could not flourish in such a society. The beginnings of a new humanitarianism were starting to take root, however. About this time the Royal Humane Society was interested in the resuscitation of drowned persons. The Slave Trade was abolished in Britain in 1807. Thomas Wakley founded the medical

journal, *The Lancet,* and campaigned for the reform of hospital management and of the training system for doctors.

    Early efforts to control the pain of surgical operations met with little success. Henry Hill Hickman, a country practitioner from Ludlow in Shropshire experimented with the effects of carbon dioxide on animals, but his "Letter on Suspended Animation" met with little response. Hypnotism had

Flogging was common in the Navy and the conditions of seamen generally showed the brutal attitude of the officers. The men quickly became cruel, if they had not been so before (*left*).
A substance used as an anaesthetic before the discovery of chloroform was alcohol. Patients were often given a great deal to drink, but in this case the fumes are being used (*above*).

a significant following, the main advocates being John Elliotson in Britain and James Esdaile in India, who published their results in the 1840s. Their work, however, was to be supplanted by the introduction of ether and chloroform shortly afterwards. Real progress was to occur in the New World. Here on the frontier of civilization was found the necessary impetus for the birth of anaesthesia. William E. Clarke and Crawford Williamson Long had both used ether in 1842, but neither published his results at the time.

In 1844 came the first indication of a breakthrough. Horace Wells was a dentist in Hartford, Connecticut. On 10th December he attended a demonstration of the effects of laughing gas (nitrous oxide) inhalation. This was organized as a public entertainment by one Gardner Quincy Colton. Wells noticed that one member of the audience who inhaled the gas injured his leg without apparently feeling pain. Next day he persuaded Colton to administer the gas to him while a tooth was extracted by Riggs, another dentist. Wells felt no pain, and after this experience used it on patients in his practice. Encouraged by success, he obtained permission to demonstrate the method at the Massachusetts General Hospital in Boston. Here, however, the patient cried out, the audience was not impressed, Wells was ridiculed, and the discovery of anaesthesia was stillborn.

It is to Morton that posterity has awarded the main credit for the introduction of anaesthesia. Morton was of different temperament to Wells. He was determined to find an answer to the pain of surgical operations. Also a dentist from the environs of Hartford, Morton enrolled as a medical student in the Harvard Medical School in Boston. Here he became acquainted with Dr Charles T. Jackson who suggested to him the use of ether.

Morton did not make the mistake of Wells by

rushing headlong into clinical demonstration. He spent much energy and many hours on experimentation, at first with goldfish and later on his dog before cautiously inhaling it himself. On 30th September 1846 he administered ether to a patient, Eben Frost, without pain. The very next day he applied for a patent, hoping to keep the nature of the agent secret. Finally Morton arranged to administer ether to a patient at the Massachusetts General Hospital. The operation was arranged for ten o'clock on 16th October. The patient, Gilbert Abbott, was ready. At a quarter past ten, Morton

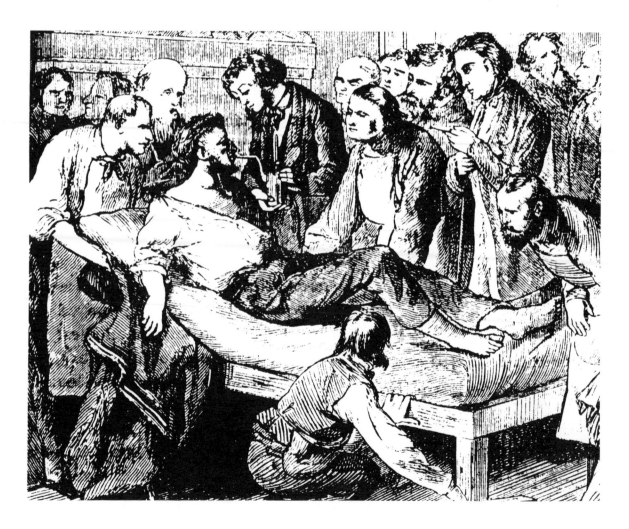

had not arrived and the surgeon, Dr Warren, prepared to start without him. Just in time Morton arrived. The administration commenced. The surgery was carried out with no appreciation of pain. Anaesthesia was born.

The Cunard wooden paddle steamer, *Acadia*, arrived in Liverpool on 16th December 1846 from America. It carried tidings of the discovery of ether anaesthesia to Europe. Dr Warren and Dr Bigelow had sent copies of the *Boston Journal* which carried the news in its issue of 18th December, and Dr Bigelow had written a personal letter to his friend

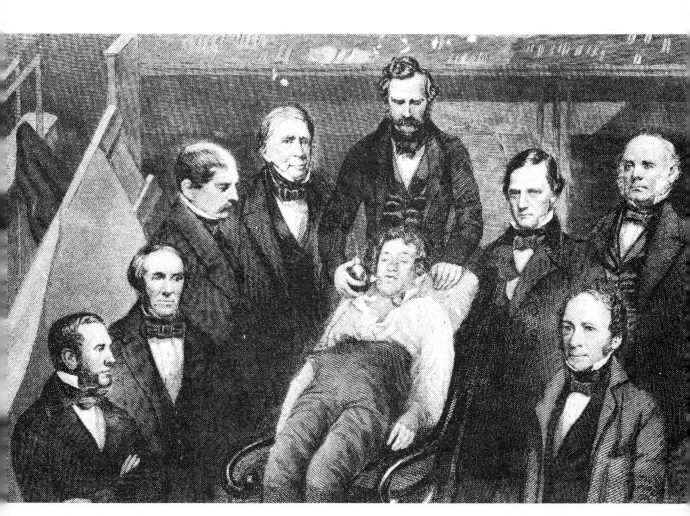

William Morton administering ether as an anaesthetic during an operation in Massachusetts in 1846 (*facing page*).

The first public administration of ether. The patient is Gilbert Abbott, Morton is behind his head and Warren, the surgeon, is on Morton's right (*below*).

The contrast between this operation, performed without anaesthesia, with the patient blindfolded while the priest reads prayers, and the operation shown on page 33 is most striking (*right*).

Dr Boot. Dr Boot, born in Boston of a Scottish mother, had graduated in Edinburgh, but practised in London, and had made numerous visits to America. He lived in Gower Street, London, and immediately on receipt of the news assisted in the extraction of a molar tooth from a Miss Lonsdale with the help of the new ether anaesthesia. He communicated the news to the surgeon, Robert Liston, who had also originated in Edinburgh.

On 21st December 1846 Liston amputated the

leg of a butler, Frederick Churchill, while Squire, an instrument maker from nearby Oxford Street, administered ether. Surgery in the days before anaesthesia had to be speedy. Liston took only 28 seconds to complete the operation. This successful demonstration in University College Hospital impressed the audience assembled and was the signal for the success of ether in London. Liston himself epitomized the fact in his famous remark, "This Yankee dodge, gentlemen, beats mesmerism hollow." The medical journal, *The Lancet*, reported with enthusiasm: "that its discoverer should be an American is a high honour to our American brethren; next to the discovery of Franklin, it is the greatest contribution of the New World to Science."

When the *Acadia* docked in Liverpool, the ship's doctor, Dr Fraser, set off for his home in Dumfries. He took the news with him of the discovery of anaesthesia and on 19th December, the same day that Dr Boot used ether in London, Dr Scott used ether in the Dumfries and Galloway Infirmary.

Scott was a young man of twenty-six. He had gained his MD from Edinburgh University and practised in Dumfries as a surgeon, earning the nickname "Butcher Scott" as being much given to the use of the knife. Nevertheless, here was another Edinburgh graduate who was quick to appreciate the significance of the ether discovery and to apply it to his practice. In 1872 he wrote to the editor of the *Lancet* to establish his prior claim over Liston: "My much esteemed and lamented friend the late Sir J. Y. Simpson, having investigated the facts, with the statement I have made, was so satisfied with the authenticity of it that he not only in his lectures to the students attending his class, but also in his lectures on Anaesthetics delivered before the Royal College of Surgeons in March 1868, stated the priority of my claim to Mr Liston." The Dumfries and Galloway Royal Infirmary was the fourth

Thomas Wakley, the first editor of the *Lancet*, which started as a crusading paper, publicly criticizing incompetent doctors. The *Lancet* warmly welcomed anaesthesia (*above*).

It was a paddle steamer like this which carried the news of Morton's first successful administration of ether across the Atlantic. It took a long time for the news of medical discoveries in other lands to arrive (*above*).
Many brave experimenters tried their new discoveries on themselves before risking their patients' lives. This is Charles Jackson experimenting on himself with ether (*right*).

voluntary hospital to be built in Scotland. Sadly, the old building has since been demolished.

Simpson himself heard of the discovery of anaesthesia from London. He soon went to London to see for himself. The Christmas vacation of 1846 was spent there as Liston's guest. Returning to Edinburgh he used it on 19th January 1847 for the operation of craniotomy in a case of obstructed labour.

London was the centre of the British Empire and the hub of the communications network of the time. News of the action of ether spread rapidly around the civilized world. It took only a little over five months to reach Australia. In those days sailing ships took several months to cross to the other side of the globe. The *Lady Howden* left London on 29th January 1847 and berthed at Hobart Town in Tasmania on 27th May. Ether anaesthesia was administered in St John's Hospital, Launceston, on 7th June by Dr William Russ Pugh. Pugh communicated the news to the Australian Medical Journal in a short paper written the same day. Pugh later moved to Melbourne and finally came to Britain. He died in London on 27th December 1897 at the age of ninety-one and his tomb may be viewed today in a Brighton cemetery.

Amputations without an anaesthetic, like this one in Tudor times, meant that surgery had to be fast. Most of the very intricate operations performed today would be impossible without anaesthesia.

4 Chloroform

As had the discoverers of ether, Simpson experimented on himself with chloroform.

Simpson's new house, 52 Queen Street.

Simpson was quick to appreciate the advantages of ether in his obstetric practice. Following his first administration of it on 19th January 1847, after which the mother made an excellent recovery, he was encouraged to use the agent freely. In other countries ether was also being used in midwifery. In France it was given for a difficult forceps delivery on 30th January. In America Dr Walter Channing used it independently. To Simpson, however, is given the credit for the extension of the use of anaesthesia to the obstetric field. He practised and preached its use, and he was the first to publish his results. A paper in March 1847 was widely circulated and as a result he was made a member of the Berlin Obstetric Society.

Nevertheless, ether was not entirely satisfactory. It had an unpleasant smell, was irritating to the eyes, and its use tended to be followed by nausea and vomiting. Simpson began to look round for something better.

He had moved into a new house, 52 Queen Street, in 1845. This was a larger house than he had previously occupied. There was room for his growing family, his servants, and his increasing practice. It was situated conveniently near to the centre of Edinburgh and was to be his home for the rest of his life. It was in this house that Simpson began to experiment with the inhalation of various substances. Inhalations were usually carried out in the late evening or early morning. Professor Miller, the surgeon, was a close neighbour, and he took an interest in these experiments and would often look in to see how the researches were progressing.

Simpson was not a scientist and had no experimental model except himself. What risks he took in the inhalation of various vapours can only be conjectured. On at least one occasion when Dr Guthrie recommended a particular volatile liquid, Lord Playfair, the Professor of Chemistry, insisted that the substance be tried first on two rabbits, and both rabbits died!

In October 1847 David Waldie, a student contemporary of Simpson's, who had given up medicine and turned to pharmaceutical chemistry in Liverpool, visited the Simpson household. Waldie was familiar with chloroform which had been prepared in 1831 and whose physical and chemical properties had been described by the French scientist, Dumas, in 1835. Flourens had already shown that it had an anaesthetic effect on animals, but this fact was not known to Waldie and Simpson. Waldie promised to supply a quantity of chloroform to Simpson for trial use, but he was unable to deliver it. Some was eventually obtained from the local chemists, Duncan and Flockhart, but the liquid appeared heavy and unpromising, and it was at first put aside.

Dr George Keith and Dr J. Matthews Duncan were assistants to Simpson at this time. They helped in the researches. Matthews Duncan once visited a laboratory and took away samples of every liquid which appeared capable of being inhaled. Such was the crude nature of their experimental approach.

The crucial experiment was carried out on the evening of 4th November 1847, in Simpson's dining room, the main ground floor front room at 52 Queen Street. James Simpson, Matthews Duncan and George Keith sat down and began to try the inhalation of various liquids which they had collected. At first they had little success, but then it was decided to try the chloroform. The bottle was searched for and eventually recovered from beneath a pile of waste paper. Glass tumblers were charged and the inhalation begun. After a time the three became insensible. Simpson awoke to find himself

Two pictures of the room in which chloroform was discovered in the Queen Street house. Among Simpson's own possessions in these photographs are the clock on the mantelpiece, vegetable dishes, Jacobean sideboards, stethoscopes and the pill box.

1848

ACCOUNT OF A NEW ANÆSTHETIC AGENT, AS A SUBSTITUTE FOR SULPHURIC ETHER IN SURGERY AND MIDWIFERY*

By J. Y. SIMPSON, M.D., F.R.S.E.,

PROFESSOR OF MIDWIFERY IN THE UNIVERSITY OF EDINBURGH; PHYSICIAN-ACCOUCHEUR TO THE QUEEN IN SCOTLAND, ETC.

I esteem it the office of the physician, not only to restore health, but to mitigate pain and dolors.—BACON.

Communicated to the Medico-Chirurgical Society of Edinburgh at their Meeting on 10th November, 1847

NEW YORK:

REPUBLISHED BY RUSHTON, CLARKE AND CO., CHEMISTS AND DRUGGISTS, 110 BROADWAY, AND 10 ASTOR HOUSE,

1848

THIS REPRINT IS DEDICATED,

WITH RESPECT,

TO

THE MEDICAL FACULTY OF THE UNITED STATES

BY THEIR OBEDIENT SERVANTS,

RUSHTON, CLARKE AND CO.

The Title Page of the American edition of Simpson's account of his new anaesthetic agent.

on the floor, Duncan beneath a chair quite unconscious and snoring, and Keith kicking the legs of the supper table. After an interval all three recovered and expressed themselves delighted with the experience. The inhalation was repeated many times, and one of the ladies, a Miss Petrie, a relation, also participated. Finally, at about three o'clock in the morning, supplies of chloroform were exhausted and the party broke up to retire to bed.

All those present had been impressed. These included the ladies of the family and one Captain Petrie. Captain Petrie had led an adventurous life which included naval service in the Caribbean against the French, and he had led a detachment of men in the burning of the Capitol in Washington in 1814. Now he had been privileged to observe a historic event which was to have even greater significance for the future of mankind.

Simpson was now sure that he had at last found an anaesthetic agent superior to ether. On 8th November he first used it for an obstetric case. The patient was a doctor's wife, Jane Carstairs. Simpson rolled a pocket handkerchief into a funnel shape and moistened it with liquid chloroform. The mother's pain was relieved by the inhalation and the baby born was a girl subsequently christened Wilhelmina, on Christmas Day, 1847. This child grew up and kept in touch with the great obstetrician. When she was seventeen years old she was photographed and Simpson kept her picture above his desk, jokingly referring to her as "Saint Anaesthesia."

Simpson was not slow to communicate his new discovery to his medical colleagues. He read an account to the Edinburgh Medico-Chirurgical Society on 10th November, and a written pamphlet appeared on the 15th. His classic paper, "On a New Anaesthetic Agent, More Efficient than Sulphuric Ether" appeared in *The Lancet* on 20th November.

The next step was to try chloroform in surgical

45

operations. Professor Miller, Simpson's neighbour who had been so interested in the experiments at No. 52, was keen to cooperate. The opportunity arose when a patient with strangulated hernia was brought into the Royal Infirmary on 9th November. Simpson, however, was not available and the operation had to proceed without anaesthesia. The patient collapsed soon after the incision was made and did not recover. Had he been receiving chloroform such a death might have been attributed to the new method and the advance of anaesthesia halted.

Another opportunity presented itself next day. Miller was to operate on the arm of a six-year-old child. A crowd assembled in the operating theatre to watch the event. The child was from the west of Scotland and spoke only Gaelic. Simpson could not converse with the screaming, fighting child, but succeeded in holding the chloroformed handkerchief to his face until he relaxed into a state of unconsciousness. The operation was carried out expeditiously, a piece of dead bone being removed, and the child awoke later having felt no pain or awareness of the operation.

Two other surgical patients received chloroform that day. It is of interest to report that Professor Dumas, Dean of the Faculty of Science in Paris, and original describer of the properties of chloroform, was among those who witnessed the historic administrations. By chance, he was visiting Edinburgh at this time, and Simpson was pleased to point out the connection between chloroform anaesthesia and the distinguished visitor.

Chloroform seemed to have such advantages over ether that in Britain the latter was soon relegated to a secondary position. Chloroform was more pleasant to inhale and was so potent that the patient was soon ready for the surgeon. The method of administration was simple and straightforward. It is no wonder that Simpson was hailed as a great pioneer.

In the public mind the discovery of chloroform became synonymous with the discovery of anaesthesia. Simpson himself was an established practitioner with authority and enthusiasm. He was already a medical writer of distinction and had ready access to the professional journals. Within a short time the use of chloroform spread round the world. In London it was given at St. Bartholomew's Hospital on 20th November. It was in use in India by September 1848.

Simpson's success in Edinburgh can be contrasted with Morton's failure in America. Although he had triumphed in his demonstration at the Massachusetts General Hospital in 1846, the subsequent history of anaesthesia in the New World was marked by controversy. Much wrangling took place as to who should be given credit for the discovery of anaesthesia. Morton petitioned the US Congress for recognition, hoping for financial reward, but his claim was challenged by Jackson. Though Russia and Sweden honoured Morton, other countries recognized the claims of Jackson. At home Morton was not in his lifetime recognized as the pioneer of ether anaesthesia. In the American Civil War he served on the side of the North, administering ether to the wounded soldiers. Another rival claimant for honour, Crawford Long, was head of the Confederate military hospital at Athens, Georgia, also using ether in the treatment of the wounded. Morton died a few years later on 15th July 1868, at the early age of forty-eight. While driving across Central Park in New York he suffered a cerebral haemorrhage and was taken dying to a nearby hospital. He died a disappointed man. Though recognized by the attending physician as one who had done "more for humanity and the relief of suffering than any man who has ever lived," his wife could only display three medals, the sole recompense he had ever received.

This 18th century apothecary's advertisement shows how far pharmaceutical chemistry overlapped with medicine proper at that time.

Simpson on the other hand was to achieve fame and honour. In 1847 he was appointed Physician Accoucher to the Queen for Scotland. In 1866 he was to become a Baronet and the University of Oxford was to make him a Doctor of Civil Law. These and many other honours were to come in recognition of his considerable achievements.

The Manufacture of Chloroform

In the first half of the nineteenth century the distinction between apothecaries, druggists and pharmaceutical chemists was not clearly drawn. There was an overlap between the medical and pharmaceutical professions. For example, a boy could become an apprentice to learn the trade, or a medical man could abandon practice to take up the chemists' trade. Such a one was David Waldie, once a licentiate of the Royal College of Surgeons and later chemist to the Liverpool Apothecaries Company. It was Waldie who had suggested the use of chloroform to Simpson but had been unable to supply him with a sample because of a fire in his laboratory. However, there were manufacturing firms in Edinburgh. We are not sure how Simpson came to consult the firm of Duncan and Flockhart, but there is a strong tradition that he first approached Thomas Smith of T. and H. Smith Ltd. Thomas Smith, however, was busy with other matters that day and recommended him to try Duncan and Flockhart who were in partnership in 52 North Bridge, Edinburgh.

John Duncan was born at Kinross on 26th August 1780. He served an apprenticeship in Edinburgh and then spent some time in London with a firm with premises near Leicester Square. In 1806 he set up a shop of his own in Perth. He adopted the new method of dispensing pills in boxes instead of paper, and ointments in pots instead of mussel shells. After initial opposition from the conservative members of the town he became successful and noted for his teaching of apprentices. At the age of thirty-eight he took a man named Ogilvie into partnership and the business became known as "Duncan and Ogilvie." Finally Duncan decided to move to Edinburgh, fitting out a large shop at 52 North Bridge. Ogilvie remained in Perth and Duncan later singled out Flockhart as a young man of excellent promise. Flockhart, born in Kinross-shire in 1808, was one of the first apprentices when the shop was opened at North Bridge and in 1830 became a licentiate of the Royal College of Surgeons. In 1832 Duncan took Flockhart and another able man, Anderson, into partnership, the firm then becoming known as "Duncan, Anderson and Flockhart". Anderson soon left; he became a surgeon, joined the Turkish Army and was killed. Thus in 1833 the business name again changed to "Duncan and Flockhart," later "Duncan, Flockhart and Co.," and this name was to become well known to all medical practitioners in Britain until quite recent times.

Duncan and Flockhart were ideal partners. Though different in temperament they worked well together and both were founder members of the Pharmaceutical Society of Great Britain which was founded in 1841. Flockhart worked hard in the business at North Bridge. The shop was open for fourteen hours a day. Duncan bought a farm on the south side of the city and became interested in the cultivation of medicinal plants.

Duncan and Flockhart were always ready to oblige their medical friends, and had no difficulty in preparing samples of chloroform for Simpson's experiments. After the introduction of chloroform anaesthesia they were kept busy manufacturing further supplies and became known for this product all over the world. The Crimean War provided an

> Dear Sir,
>
> In answer to your request allow me to state that I have tried different varieties of chloroform made by different manufacturers; but I have always preferred yours because I have always found it the best. — I have now used several thousand ounces of the article manufactured by you; & never saw in any single case any bad effect that I could attribute to its employment. Perhaps I may add that I have seen specimens of chloroform so bad that I would be terrified to exhibit them to any patient; & one of these specimens was from a large & influential London house, & of their own manufacture.
>
> Yours very truly
>
> J Y Simpson
>
> To Messrs Duncan
>
> Edinbgh 22 Oct - 1847

A letter from Simpson to Duncan and Flockhart and Co., recommending the chloroform they manufacture (see p. 52).

Florence Nightingale's work in the Crimea was vastly benefited by the relief of pain during operations.

additional impetus to the demand for chloroform. Duncan and Flockhart were able to meet these requests and the further orders brought about by the Franco-Prussian War of 1870. The quality of the product was high. The chloroform bottles supplied carried the name of the firm as a guarantee of purity—one of the first pharmaceutical products to do so. Simpson himself wrote commending the chloroform of Duncan and Flockhart as the best available.

Hannah Greener was the first patient to die under chloroform (see p. 54).

John Duncan died in 1871 at the age of ninety-one. Flockhart had died some three weeks earlier at the age of sixty-three. The fortunes of the firm continued to prosper, however. Connections with the Simpson family continued when John Simpson, a nephew of Sir James, was admitted as a partner in 1863, though the association was to be cut short by his early death at the age of thirty-nine. More recently, Dr W. T. Simpson, great grandson of Alexander (Sandy), James Simpson's older brother, has been medical adviser to the firm. Also, in recent years, a series of amalgamations have resulted in the old firm becoming submerged as part of a greater organization, British Drug Houses Ltd., and, sadly, after many years, the household name of Duncan and Flockhart has ceased to exist.

The Safety of Chloroform

The fight for anaesthesia was not over. The discoveries of 1847 spread widely, but not all medical men were as enthusiastic as Liston. He wrote from London that he much preferred chloroform to ether and was now using it constantly. There were opponents who felt that their basic beliefs, principles and prejudices were being challenged.

Chloroform was opposed on medical grounds. It was argued that the use of anaesthesia would increase the already great mortality of surgical operations. Simpson undertook an extensive statistical investigation to disprove this. He compared the results of amputations of the thigh in hospitals where anaesthesia was not used with those when the patients had received chloroform. A death rate as high as sixty-two per cent in Parisian hospitals and not lower than thirty-six per cent in Britain, when chloroform was not employed, could be set against a seventy-five per cent survival rate in a series of amputations using anaesthesia. The latter series was certainly small, but it was enough to point the way.

The danger of chloroform anaesthesia was always present, however. The first recorded death occurred on 28th January 1848, when Hannah Greener of Winlaton in County Durham died under chloroform at the age of fifteen years. Dr Meggison of Whickham, a nearby village, was the anaesthetist while his assistant, Mr Lloyd, was going to remove an ingrowing toe nail. Death was quite sudden as the patient sat up to inhale chloroform. Hannah Greener was buried on 30th January. The historic entry in the Register of Burials for the Parish of Winlaton shows that the funeral was conducted by the Rev Charles Tinley, though the marginal entry, "died from effects of chloriform," indicates a lack of familiarity with the correct spelling of this new agent's name.

This dangerous ability of chloroform to cause sudden death, even though the patient was not deeply anaesthetized, was apparent only on rare occasions. Many practitioners were able to show relatively large series of administrations without harm to the patients. But sporadic reports of chloroform deaths continued. The Edinburgh method of using the drug was held to be particularly safe by its advocates. They preached simplicity of administration: putting a towel moistened with chloroform over the nose and mouth and paying careful attention to the patient's proper respiration. If the patient's breathing was satisfactory, it was taught, the heart would look after itself. Anaesthetists today would agree that the most important thing was to make sure that the oxygen in the air reached the patient's lungs, but would regard the teaching of Simpson and the Edinburgh School as something of an oversimplification. John Snow held other views. The leader of anaesthesia in London, he believed that it was necessary to regulate the concentration of chloroform inhaled very carefully and taught the more scientific approach to the subject. Simpson himself had to wait many years for a personal experience with chloroform death. It was not until February 1870, some few months before his death, that he administered chloroform to a patient in a small village hospital while Mr Brotherton operated for an ovarian mass. As the surgeon

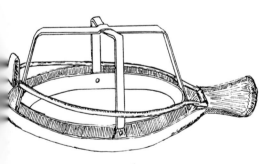

This mask, one of the earliest used for administering anaesthesia remained in use for many years. It is called a Schimmelbusch mask.

put his hand in the incision the patient vomited profusely. Simpson reported that the patient's eyes immediately opened, his pupils dilated, his face looked pallid and respiration ceased. Resuscitation was unavailing.

Today it is known that chloroform can occasionally produce ventricular fibrillation, particularly during induction of anaesthesia in a frightened patient. In ventricular fibrillation the muscle fibres of the heart contract irregularly and independently and the lack of coordination means that no blood is expelled. Circulation ceases and death results in an abrupt manner. This mechanism was unknown until 1911 when Goodman Levy showed that it was the explanation of many chloroform deaths. It is more likely to occur in the terrified patient when there is an increase in the adrenaline concentration in the blood. Adrenaline and chloroform combine to produce ventricular fibrillation in these cases.

The Australian anaesthetist, Embley, also wrote a classic paper in 1902 showing that chloroform can cause actual stopping of heart action by vagal inhibition in certain cases. Vagal inhibition is the mechanism of the common faint, when the heart slows and blood pressure falls as stimuli reach the heart along the vagus nerve. In the common faint, however, the effect is transitory: the subject falls to the ground and the blood flow to the brain is immediately resumed so that consciousness returns. Under anaesthesia, especially if the patient is in the upright or sitting position, this may not be such a benign phenomenon and death can result.

These mechanisms were, however, not understood during Simpson's lifetime. Deaths from anaesthesia were rare and no single practitioner saw more than a few cases. Though more unpleasant to inhale, ether was a safer anaesthetic agent than chloroform. Chloroform had advantages in its smoothness, potency and general applicability. Many doctors

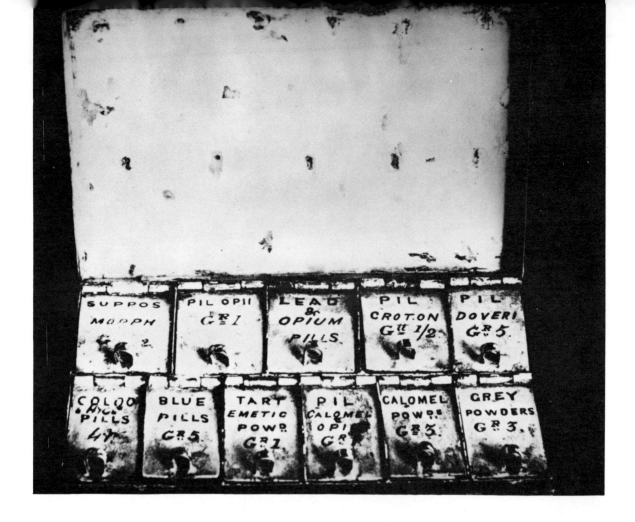

Simpson's pill box. All doctors of the time had pill boxes very like this.

felt it was the better agent. They were fortified in their opinion by a Report of the Chloroform Committee of the Royal Medical and Chirurgical Society in 1864 which confirmed chloroform's position as the best agent. The controversy continued, however, and in 1888 and 1889 the First and Second Hyderabad Chloroform Commissions reported, again stating that chloroform was not a primary cardiac depressant.

Another danger was that of possible damage to the liver. Reports of jaundice and even death due to interference with liver function following chloroform anaesthesia came in the years following 1890 and were given authority by the classic publications of Guthrie in 1894.

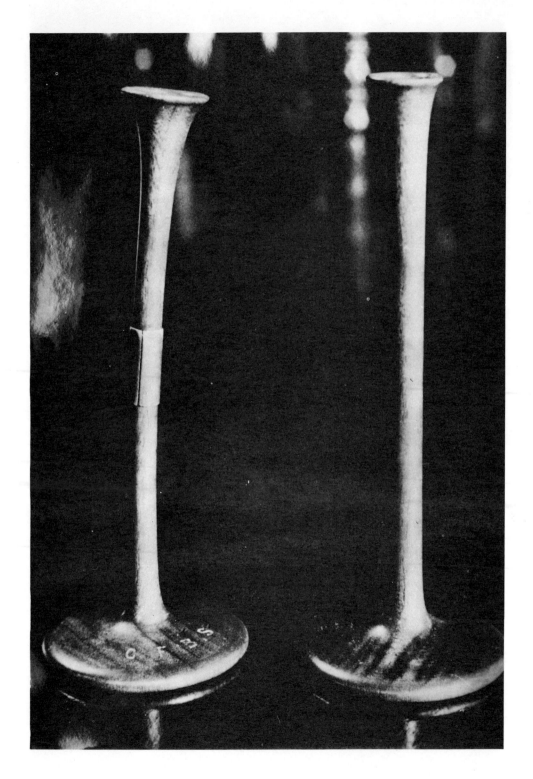

Before rubber tubes were available, stethoscopes were rigid. Many very important discoveries were made by doctors using stethoscopes like these, which belonged to Simpson.

Chloroform, which had replaced ether after its discovery in 1847, gradually came under suspicion. It became clear to many that it was not quite so safe as ether. Although death was very rare, some series of administrations having a mortality rate of between one in two thousand and one in three thousand, when it did occur the tragedy was sudden, unexpected and distressing. The record with ether was of greater safety. There was an increasing tendency to use the latter agent as a matter of choice, particularly when nitrous oxide had been reintroduced to smooth the initial breaths of the irritant ether vapour and Joseph Clover had introduced this technique and his portable, regulating ether-inhaler. Nitrous oxide was reintroduced to anaesthesia, after years of neglect, by Colton in 1863 and the technique of nitrous oxide anaesthesia in dentistry had been brought to London from Paris by the American T. W. Evans in 1868. The introduction of these methods and the increasing influence of Clover in British anaesthesia led to a falling off of the popularity of chloroform.

In 1947, as a sort of centenary celebration of chloroform anaesthesia, the well established Department of Anesthesia at Madison, Wisconsin, USA, under the direction of Ralph Waters decided to re-investigate the drug as though it were a new agent. Administered carefully with modern apparatus and methods it was concluded that chloroform still held a place in anaesthetic practice. Nevertheless, they did encounter four cases of cardiac arrest during their studies, though all these recovered fully when prompt treatment was given.

Some years later, in 1973, we find that chloroform is rarely used. Many present day anaesthetists have never used it and never seen it used. There has been such a proliferation of new agents that they do not feel the need for chloroform in their armamentarium. It is interesting, however, that halothane, another agent similar to chloroform in many respects, has become very popular and in some centres is used for the vast majority of surgical cases.

Halothane has a chemical formula somewhat similar to that of chloroform, and its pharmacological properties only differ in certain respects. Some workers in the USA have carried out a double blind study of the two drugs and been unable to distinguish between them. In a double blind study the doctor does not know which of the two agents he is using. They appear identical to the eye. The anaesthetist uses the numbered sample in random fashion and makes a series of observations. The number code is later broken and results compared. The anaesthetists were unable to distinguish between chloroform and halothane in this type of study, although other pharmacological experiments can be performed which show marked differences. For example, the effect of the two agents on the sympathetic nervous system is quite different and this probably accounts for the fact that halothane is less likely to result in cardiac arrest. Halothane has also been blamed for causing liver damage as evidenced by the development of jaundice after the operation. Although halothane has come under suspicion in this respect, its mode of effect has been shown to be quite different to that of chloroform. Halothane is only dangerous if a rare sensitivity develops, and this alleged sensitivity is so rare that study of it is extremely difficult. In fact, after many reviews of series of cases, and after many comparisons in man and animals there is still no proof that such a toxic effect exists at all. It may be that the individual has coincidentally contracted infective jaundice!

It is, however, ironic that chloroform, for so many years the most popular anaesthetic agent in Britain, has now fallen into almost complete disuse, while halothane, a near relative chemically speaking, has come to occupy a leading position. It is, however, interesting to note that halothane is also a British drug. It was synthesized in the laboratories of Imperial Chemical Industries near Manchester in 1951, and after extensive trials on animals first used on human patients in 1956.

5 Edinburgh and London

Simpson and Syme, occupied the two Chairs which enabled them to influence the practice and teaching of chloroform anaesthesia. James Syme had been appointed Professor of Clinical Surgery in 1833 and was the leading surgeon in the city. Simpson and Syme both made considerable contributions to their own specialities as well as to the new art of anaesthesia, but they were seldom able to agree on medical or medico-political matters within the University.

James Syme was born in 1799. After qualifying he embarked on a surgical career and became well known following a successful amputation at the hip joint in 1823, when a young man survived long enough for the surgeon to win acclaim. Syme took over the anatomy class from Liston, but the two men later quarrelled. Either the young man disapproved of Liston's collaboration with body snatchers, or the older man became jealous of Syme's increasing fame.

James Syme published numerous papers, and later his *Principles of Surgery* and *Treatise on the Excision of Diseased Joints*. In 1833 he was appointed to the Chair of Clinical Surgery. By a strange clause in the contract of his predecessor, James Russell, compensation was due on resignation from the life appointment. Syme as his successor had to pay him £300 a year as long as he lived, a fact which has led him to be accused of having purchased the Chair!

When Liston left Edinburgh in 1835 to become Professor of Surgery at University College Hospital in London, Syme became the leading surgeon in Edinburgh. Simpson was junior in age and the

Simpson's cross with the lancet blade open.

RESEARCHES,

CHEMICAL AND PHILOSOPHICAL;

CHIEFLY CONCERNING

NITROUS OXIDE,

OR

DEPHLOGISTICATED NITROUS AIR,

AND ITS

RESPIRATION.

By HUMPHRY DAVY,

SUPERINTENDENT OF THE MEDICAL PNEUMATIC INSTITUTION.

LONDON:

PRINTED FOR J. JOHNSON, ST. PAUL'S CHURCH-YARD.

BY BIGGS AND COTTLE, BRISTOL.

1800.

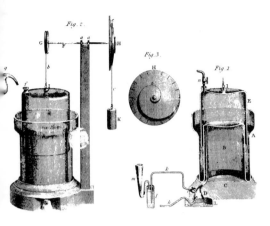

Facing Page—The title page of Humphrey Davy's book about nitrous oxide. This substance came back into favour in 1863 (see p. 59). Humphrey Davy's "gas machine" shows how crude the early anaesthetic apparatus was (*above*).

specialty of midwifery was held in less esteem than surgery. The two men were to disagree over a wide range of subjects. For example, Syme did not believe in the teaching of pathology as a separate subject, preferring that it be dealt with by the clinician in any particular discipline. Simpson, with a more progressive outlook, wrote a memorandum in support of its being a department in its own right. In this particular dispute the younger man won. The Chair of Pathology was continued, but the new man appointed to the post, William Henderson, soon caused trouble by his advocacy of homoeopathy.

Homoeopathy had been originated by Samuel Hahnemann of Leipzig. He believed that the correct treatment of any disease lay in the use of tiny doses of drugs which tended to mimic that particular condition. Established apothecaries and physicians opposed these views. For once Syme and Simpson acted together. In the meeting of the Medical Chirurgical Society of Edinburgh Syme moved that persons who admitted that they practised homoeopathy be disqualified. Simpson seconded the motion.

The two men were very different in their personal characters. In the conduct of his practice Simpson showed a somewhat haphazard approach. He preferred to use his memory rather than keep methodical notes and accounts. Provided he had a basic minimum of money for everyday living, he was careless about monetary matters. The house was crowded with patients who were treated alike regardless of the expected fee. Lots were drawn daily to determine precedence to see the great man. Money was not asked for if the patient was in financial difficulty, but all sorts of presents from patients were delivered to the house.

Certain patients paid quite large fees, and as Simpson always had more than necessary to supply his own wants he could afford to adopt this attitude.

He made loans freely to friends in temporary distress, always remembering his early struggles and the help he had received from his family then. On one occasion it is said that Simpson's sleep was disturbed by a rattling window. In the dark he found a piece of paper to fill the chink. Only next morning did Mrs Simpson discover that he had used a ten pound note, recently received as a fee and not put in a safe place!

Few personal papers of Syme survive. He was a reticent man and his private life and thoughts were kept to himself. He lived at Millbank, in a country house at Corstorphine on the western outskirts of the town, rising early, visiting his garden and attending to his letters before breakfast, and then walking to his consulting rooms, spending the remainder of the day in the infirmary, operating or teaching. He was a precise and skilful operator without feeling the need for theatrical mannerisms. It has been said that he never wasted a word, a drop of ink, or a drop of blood.

On return from a visit to London, Simpson fell ill with a nasty abscess in the armpit, secondary to a poisoned finger. Jessie and his assistants were sure that a surgical opinion was needed and they called in Syme. Simpson himself would probably have preferred to consult Miller but the other man had already been summoned. Simpson was dangerously ill for some time but he recovered gradually and went on a continental holiday for convalescence. Simpson's lecture to the new graduates in 1842 had contained the words, "No man will in any case of doubt or danger intrust to your professional care the guardianship of his own life, or of the life of those who are near and dear to his heart, merely because you happen to be on terms of intimacy with him." In those days, before the advent of antibiotics to deal with infection, a poisoned finger with axillary abscess could easily proceed to septicemia and

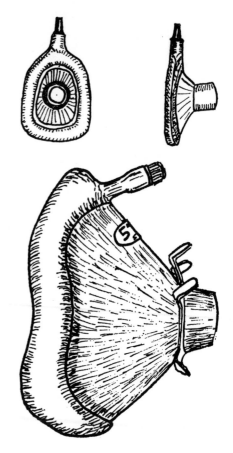

In order that the patient shall breathe only the gas supplied by the anaesthetist modern face masks fit the face closely and are produced in different sizes for adults and children. Compare these modern masks with the simple mask on page 55.

death. Jessie had echoed her husband's words when she summoned not Miller, his friend, but Syme, the surgeon with the foremost reputation in the city.

Further Developments in Anaesthesia

In Edinburgh Simpson and Syme greatly influenced the teaching and practice of chloroform anaesthesia. The latter used to give a ten minute lecture which encompassed the philosophy as then understood. Stress was laid on simple principles. No estimate was made regarding the medical fitness of the patient to receive chloroform, it being made freely available to all who required it. No attempt was made to restrict the dose of the drug used, emphasis being placed instead on the patient's response. It was considered essential to allow free admixture of air with the chloroform vapour, and to use a folded towel or handkerchief to provide a large area to receive the liquid. It was thought that the more rapidly the vapour was given the better the effect. Respiration was more important than circulation. While the practice of keeping a finger on the pulse was discouraged, stertorous breathing was thought to be of serious import. When this occurred the tongue was seized with forceps and pulled well forwards. The patient was kept horizontal, the London habit of anaesthetizing patients in the sitting position being deprecated. Syme gave full credit to Simpson for the development of the art.

In London, however, chloroform anaesthesia had evolved in a somewhat different manner and there were to be disagreements between the two schools of thought. The chief figure in London was John Snow. Born in York in 1813 he had studied at Newcastle Infirmary before migrating to London where he attended the Westminster Hospital before passing the qualifying examinations to become a member of the Royal College of Surgeons of England and of the Apothecaries Hall.

The advent of ether interested the young practitioner. Sir Benjamin Ward Richardson, Snow's biographer, gives a fascinating description of an early encounter,

> One day... he met a druggist whom he knew bustling along with a large ether apparatus under his arm. "Good morning!" said Dr. Snow. "Good morning to you, doctor!" said the friend; "but don't detain me, I am giving ether here and there and everywhere and am getting quite into an ether practice. Good morning, doctor!" Rather peculiar! said the doctor to himself; rather peculiar, certainly! for this man has not the remotest physiological idea. "An ether practice! if he can get an ether practice, perchance some scraps of the same thing might fall to a scientific unfortunate!"

Snow was already making an improved inhaler, trying it out on animals and on himself. He lost no time in applying himself to clinical practice and soon became the leading anaesthetist in London. Snow had a different approach to Simpson, the scientific approach. He gradually gave up his other medical practice to concentrate on anaesthesia, but he was interested in research as well as in clinical practice. He was also quick to publish. In September 1847, only one year after the introduction of ether, he was able to embody the whole of his work to date in his book *On the Inhalation of the Vapour of Ether in Surgical Operations*. This was the beginning of the scientific approach. But the book was hardly completed when chloroform was discovered and ether thrown into the shade. Snow soon appreciated the valuable properties of chloroform and it became the main agent he used.

John Snow was a bachelor and a man of temperate habits. He was methodical in his work and kept careful records of his chloroform administrations.

John Snow, M.D.

to some varicose veins of the leg.

Also to a young woman which he performed excision of the knee joint. This patient was through the operation very well but died the next week but one afterwards

Monday 4 April 1853

Administered Chloroform to Mrs Gatenby, 7 Seymour Place Wandsworth Road whilst Mr Fergusson operated for hemorrhoids. The patient a stout robust person, aged about 45 to 50 resisted the chloroform a good deal after she became half unconscious. She vomited after the operation, and became very livid in the face for a minute or so in consequence of holding her breath whilst vomiting

Tuesday 5 April

Administered Chloroform at Mr Juddle to a gentleman Lord Chas Battie about 35 with a bad stricture whilst Mr J. endeavoured to get a catheter into the bladder. The chloroform was continued for an hour. No sickness

Thursday 7 April

Administered chloroform to the Queen in her confinement. Slight pains had been experienced since Sunday. Dr Locock was sent for about nine o'clock this morning, strong pains having commenced, and he found the os uteri had commenced to dilate a very little. I received a note from Sir James Clark a little after ten asking me to go to the Palace. I remained in an apartment near that of the Queen along with Sir J. Clark Dr Ferguson and (for the most part of the time) Dr Locock till a little . twelve

He tried its use not only in surgical operations, but also in the treatment of various distressing medical conditions including convulsions, lockjaw, facial neuralgia and vomiting in pregnancy. A tireless worker, seldom taking any holidays, Snow became famous for his administration of chloroform to Queen Victoria—"chloroform à la reine"—for the birth of Prince Leopold in 1853 and again for the birth of Princess Beatrice in 1857. He was also well

Snow's diary for 7th April 1853, recording the birth of Prince Leopold (*facing page*).

Mayall's photograph of Queen Victoria (1861). Her use of chloroform during the birth of her children helped to remove religious objections to the use of anaesthesia.

known for his work on cholera. He made careful epidemiological studies on the London water supply and was able to demonstrate beyond doubt that cholera was a water-borne disease.

Snow's scientific mind contrasted with the easy-going approach of Simpson and Syme in Edinburgh. There is no evidence that any scientific research at all was undertaken by the latter. Snow was to criticize the Edinburgh teaching.

Syme pointed to the rarity of deaths in his practice as compared to the more numerous fatalities in London. Snow replied that "Mr Syme seems entirely to overlook the relative size and population of the two places. When these circumstances are taken into account, the mortality from this cause seems to be pretty equal." He agreed that Syme was correct in admitting plenty of air with the chloroform vapour, but went on to recommend that the ratio should be ninety-five parts of air to five of vapour. Snow designed an inhaler to ensure that this was so, but also suggested that when a handkerchief was used only fifteen minims (sixty-seven drops) should be put on it at any one time. Dilution of chloroform with wine or spirit also helped to prevent overdose. Snow pointed out that chloroform diluted about seven parts in spirit, under the name of chloric ether, had in fact been used in London before Simpson used the undiluted form. Chloric ether had in fact been given at St Bartholomew's Hospital in the spring of 1847 by Mr Holmes Coote to two patients of Mr Lawrence, a surgeon there. It is not therefore surprising that, following Simpson's work in Edinburgh, Holmes Coote was one of the first to use chloroform in London, administering it for three surgical cases on 20th November.

In one respect London pioneered chloroform anaesthesia ahead of Edinburgh. Surgery of the jaw and tongue rendered administration technically difficult, but Snow managed to use a sponge impregnated with chloroform. In one case he recorded, "She was quite insensible at the beginning of the operation, and afterwards I applied a sponge near the face whenever I could, but there was very little opportunity and although the patient was prevented from regaining consciousness, she cried out and struggled occasionally." Snow died in 1858. At the age of forty-five he had laid the foundations for the scientific investigation of anaesthesia. It has been said in recent years that nothing new has been discovered about the principles of anaesthesia without some reference to the idea being found in Snow's work. His final contribution was *On Chloroform and other Anaesthetics* which was published posthumously.

Snow was buried in Brompton Cemetery in London. His grave has been restored by anaesthetists from Britain and the United States. Syme is buried in a small graveyard of St John's Church in Princes Street, not far from the statue of Simpson. Both men have received less than their public due for their part in the chloroform story.

The Church and Chloroform

The introduction of chloroform also met with opposition on moral and religious grounds. The pain of parturition was said to be beneficial. The agony of the surgeon's knife was thought by some to confer some ill-defined advantage on the unfortunate recipient. Other discoveries, such as Jenner's vaccination, had met with ignorant criticism in their time. Chloroform was no exception.

Simpson answered his critics with their own weapons. When a Dublin obstetrician found chloroform to be an unnatural method, contrary to nature, the reply was that the carriage was useful in locomotion, for even in Dublin it was not convenient to walk everywhere using the legs provided by the

Almighty. Galen's aphorism was quoted, "*Dolor dolentibus inutile est*" ("pain is useless to the pained"). The use of anaesthesia in midwifery was said to be against the teaching of the Bible: "In sorrow thou shalt bring forth children" (Genesis 3: 16). Simpson answered by study of the original Hebrew text and showed that "sorrow" meant labour, toil or physical exertion, rather than pain.

When he wrote his classic pamphlet, "Answer to the Religious Objections Advanced Against Employment of Anaesthetic Agents in Midwifery and Surgery" it was headed by two texts from the New Testament: "For every creature of God is good and nothing to be refused, if it be received with thanksgiving" (Timothy 4: 4); and "Therefore to him that knoweth to do good and doeth it not, to him it is sin" (James 4: 17). The Bible could also be used to supply the most powerful answer of all: "And the Lord God caused a deep sleep to fall upon Adam; and he slept; and He took one of his ribs and closed up the flesh instead thereof" (Genesis 2: 21). Simpson further showed that the word translated into "deep sleep" might be more exactly given as "coma" or "lethargy". In these ways the fight for anaesthesia was carried into the very camps of his opponents.

John Snow also had a part to play. On 7th April 1853, he administered chloroform to Queen Victoria on the occasion of the birth of Prince Leopold. The famous event was recorded in his diary:

> I received a note from Sir James Clark a little after ten asking me to go to the Palace . . . at twenty minutes past twelve by a clock in the Queen's apartment I commenced to give a little chloroform with each pain, by pouring about fifteen minims by measure in a folded handkerchief. The first stage of labour was nearly over when the chloroform was commenced. Her Majesty expressed great relief from the application, the pain being very trifling during the uterine contractions, whilst between the periods of con-

A page from John Snow's diary for 14th April 1857, recording the birth of Princess Beatrice. Note that Prince Albert had already given some chloroform to the Queen (lines 6–7).

Tuesday 14 April.

Administered Chloroform to her Majesty the Queen in her ninth confinement. The labour occurred about a fortnight later than was expected. It commenced about 2 A.M. of this day, when the medical men were sent for. The labour was lingering and between a little after 10 Dr Locock administered half a drachm of powdered ergot, which produced some effect in increasing the pains. At 11 o'clock I began to administer chloroform. Prince Albert had previously administered a very little chloroform on a handkerchief, about 9 or 10 o'clock. I poured out 10 minims of chloroform on a handkerchief folded in a conical shape for each pain. Her Majesty expressed great relief from the inhalation. Another dose of ergot was given about twelve o'clock and the pains increased somewhat about twenty minutes afterwards. The Queen at this time kept asking for more chloroform, and complaining that it did not remove the pain. She slept, however, sometimes between the pains. Before one o'clock the head was resting on the perineum and Dr Locock wished the patient to make a bearing down effort, as he said that this would expel the birth. The Queen, however, when not unconscious of what was said, complained that she could not make an effort. The chloroform was left off for 3 or 4 pains as the royal patient made an effort which expelled the head, a little chloroform being given just as the head passed. There was an interval of several minutes before the child was entirely born; it however cried in the meantime. The placenta was expelled about ten minutes afterwards. The Queen's recovery was very favourable.

traction there was complete ease.... The infant was born at thirteen minutes past one by the clock in the room (which was three minutes before the right time) consequently the chloroform was inhaled for fifty-three minutes. The placenta was expelled in a very few minutes, and the Queen appeared very cheerful and well, expressing herself much gratified with the effect of the chloroform.

If the Queen of England accepted chloroform anaesthesia in childbirth, who was to refuse? This occasion as much as any argument made chloroform respectable. Later in the same year, on 20th October, Snow was called to Lambeth Palace for the confinement of the daughter of the Archbishop of Canterbury, and chloroform was administered. Once more he was asked to attend the Queen, on 14th April 1857, when Princess Beatrice was born. The acceptance of chloroform by society was complete, the fight for anaesthesia in childbirth won.

Simpson's own religious beliefs had been ingrained by a strict upbringing and he became a regular member of his church. He lived in a time of religious ferment marked by the Disruption from the National Church of Scotland in 1843. Intelligent men were disturbed by the scientific discoveries of the time. Geology questioned the simple chronology of the Old Testament. The theory of evolution challenged established doctrine.

When the separatist movement led to the formation of the Free Church, Simpson was one of those who joined the rebels. Pressed to speak from pulpits and at public meetings, he produced many addresses and pamphlets. In his personal life he tried to follow the precepts he preached and to embody them in his practice and teaching.

Punch shows another use for chloroform: to help in crime. This cartoon appeared in 1851.

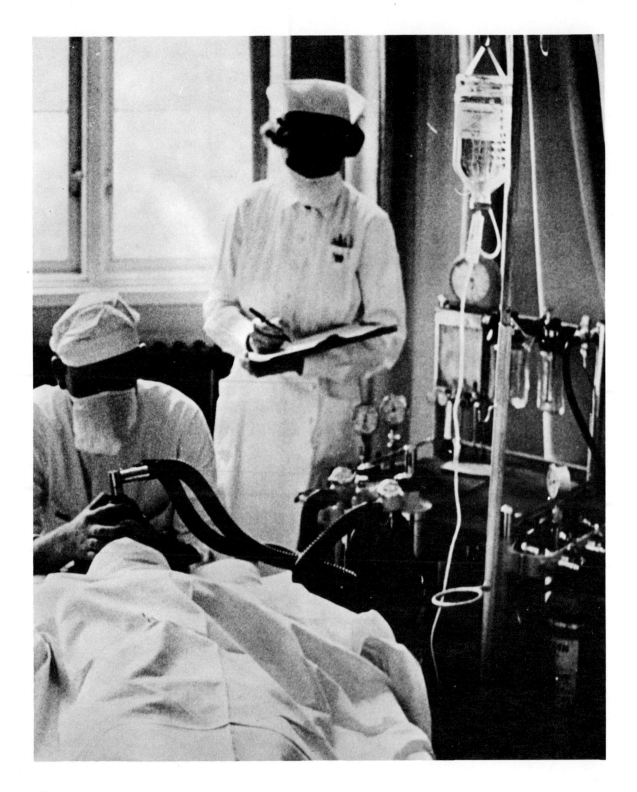

6 *Home and Abroad*

Simpson's domestic life was closely bound up with his professional activities. The practice was conducted from 52 Queen Street, and the house was busy with patients, distinguished visitors and the like. The Simpson family and servants were also numerous. There were nine children, though James and Jessie were several times distressed by bereavement. The first child Margaret, born in 1840, died at the age of four years before the move to Queen Street. She had been a source of great joy to her parents. In 1841, writing to his father-in-law, James had spoken of her, "Maggie has just been before tea laughing and skirling at a great rate. She tries to walk by a hold, and appears to think herself mighty clever when she manages a few steps." When she was taken ill the loss was sudden:

> My own dear Maggie was taken from me this morning between nine and ten o'clock. She was attacked with measles a fortnight ago, and was subsequently seized with a very bad form of sore throat, which, after several days' struggle, at last became worse on Friday night and proved too strong and fierce for her little emaciated body. It was ultimately so heart-rending to witness her terrible anxiety and restlessness, that her demise was almost a release to all of us... She asked for a "drink of water" for her little parched and burning throat a very short time before she died, but then, and for hours before, was unable to swallow it.

It appears that Maggie died from diphtheria, then a common and frequently fatal disease, there being no immunization programme nor specific treatment at that time.

The eldest son, David, was born in 1842 and grew up to qualify in medicine. His father sent him on a continental tour of the medical centres and he

It is strange to think that until Morton and Simpson, there were no anaesthetists at all.

visited Prague and Berlin. On returning home he joined his father's practice and soon made his influence felt. Not only was he popular with the patients but he also began to organize the practice in an orderly fashion. Appointment books were instituted and loose pages of notes were filed. Then, suddenly, on the threshold of success, David became ill and died. It seems that he developed appendicitis. In those days surgical removal of the inflamed appendix had yet to be pioneered. At the age of twenty-four the promising young doctor was taken away. A month later, Jessie, at the age of seventeen also went to the grave. The family fortunes seemed low.

Yet at the same time James had to consider acceptance of the highest honour yet paid to him. In January 1847 he had been appointed one of Her Majesty's Physicians for Scotland. On two occasions he had refused a knighthood. Now the Crown offered him a baronetcy. This was a great honour. Simpson was the first medical man practising in Scotland to receive this accolade. He accepted, though with some reluctance. Congratulations came from far and wide. Writing to his son, Wallie, James said:

> I have shaken hands daily till my arm is weary and sore. The proudest of all about the gift is your uncle in Bathgate. At all events, he is a thousand times prouder of it than I am. In fact, when the gift was first offered me I was rather ashamed to speak of it and doubted about accepting it. But at last it was decided otherwise.

Walter, or Wallie as he was called, was born in 1843 and lived to succeed to the baronetcy. Another daughter, Mary, died in infancy in 1847. Jessie, born in 1849, died shortly after David, as we have noted.

James, born in 1846, led an invalid's life and died in 1862. William, born in 1850, Alexander, born in

1852, and Eve, the youngest, born in 1854, all survived to adult life. The last named was a lady of character and wrote a biography of her father as well as other books. She lived until 1919. Walter succeeded to the baronetcy. He had studied the law, but lived the life of a country squire, writing a volume on *The Art of Golf*. When he died in 1898, his son James became baronet, but then the direct line ceased. There were no more descendants except through the female line.

Simpson's family life, apart from the bereavements, was a happy one. The house was always full of activity and guests were always welcome. Notable visitors included Spencer Wells the surgeon, Dumas the discoverer of the properties of chloroform, Hans Anderson of fairy tale fame, Retzius the Swedish obstetrician and Channing from Boston.

Simpson was one of those who encouraged women in the pursuit of a medical education. In this he was ahead of his time. For a time Emily, the sister of the pioneer woman doctor Elizabeth Blackwell, lodged in his house as an assistant.

The Simpson family enjoyed the company of dogs. Some would accompany their master on his rounds, riding in the carriage and sometimes going up to the bedside. The names and ages of the dogs were tabulated in a book kept next to the family Bible. One of the favourites was Puck, a black and tan terrier that lived at 52 Queen Street for fourteen years.

To escape the constant interruptions at the Queen Street house, Simpson bought a country house in the village of Trinity overlooking the Firth of Forth. The house was called Viewbank and was surrounded with roses, lilac, laburnum and apple trees planted by the new owner.

Simpson took a serious interest in archaeology as an escape from medicine. He made contributions to the literature on subjects such as the Cat-stane, a massive block of stone in a field near the Linlithgow Road, aiming to show that it might mark the tomb of Vetta, grandfather of Hengist and Horsa. Another subject he took was "Was the Roman Army provided with Medical Officers?" No historian had shown the existence of a medical branch in the Roman Army, but Simpson found inscriptions referring to army surgeons or *medici* with degrees of rank such as *medicus legionis* and *medicus cohortis*. In 1861 he was president of the Society of Antiquaries of Scotland. He was also made a Professor of Antiquities to the Royal Academy of Scotland and he was elected a member of foreign societies in Copenhagen, Athens and Nassau. Viewbank was well placed to serve his antiquarian interests, being conveniently close to the islands of Inchcolm and Inchkeith in the Firth of Forth. The former housed its hermit cell and the latter had marked stones.

Simpson often travelled further afield and was an enthusiastic user of the spreading railway service. He had visited London and the continent in his younger days. Later he was to make long journeys across the country to see patients for whom his opinion had been requested. Holidays were a rare luxury and the Isle of Man seems to have been a favourite place to visit. Often a few days relaxation and the study of some local archaeological interest would be combined with a professional visit.

Commercial ventures always interested Simpson. But they were seldom successful. The most serious misadventure concerned shares in a sugar plantation in Tobago. It was a failure and Simpson had to meet his share of the loss, a matter of several thousand pounds. Other similar excursions in the business world were to come to naught.

In 1848 Simpson was invited to join the staff of St Bartholomew's Hospital, London. There were precedents for such a move. Liston had left for London in 1834 and Syme took Liston's place for a

Because anaesthesia has made speed less important in surgery, intricate operations like this one on a lung are now possible.

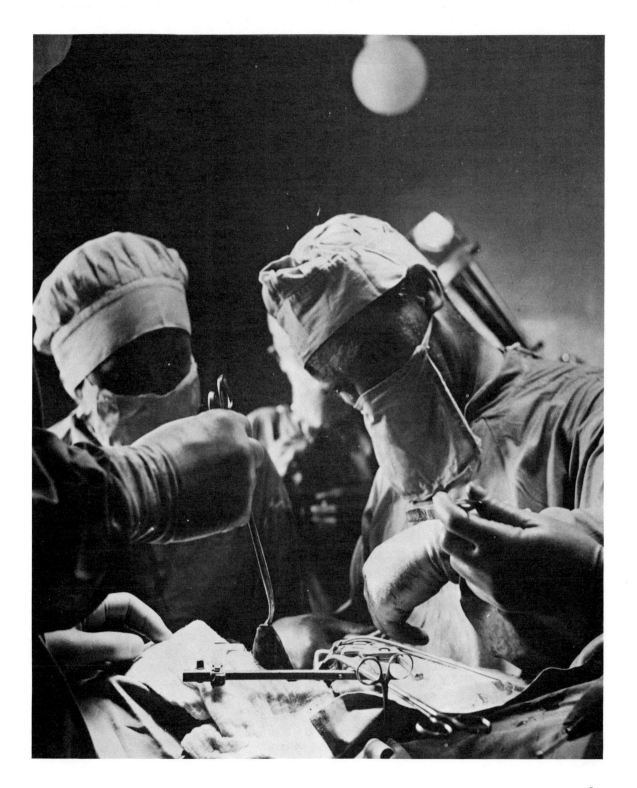

short time before returning to Edinburgh. But Simpson preferred to stay in the Scottish capital.

In his final years an attempt was made to have Simpson appointed Principal of the University, but the appointment in the event went to Sir Alexander Grant by a majority of one vote. However, the Town Council conferred the freedom of the city on him. The University of Oxford awarded him the honorary degree of Doctor of Civil Law. From Ireland came an honorary Doctorate of Medicine and an honorary Fellowship of the King's and Queen's College of Physicians of Ireland. The veteran professor was a distinguished figure, respected throughout the civilized world.

7 Simpson's Later Years

It is for chloroform that Simpson is chiefly remembered today. But throughout his life he took an active interest in all matters concerning his chosen profession, midwifery.

Long research led to his advocacy of acupressure. This was a method for the arrest of haemorrhage which was simple and elegant and did not give rise to suppurating wounds which were so common as a result of traditional sutures. This method he claimed would accelerate wound healing. Simpson believed that the technique would revolutionize surgery, perhaps even more so than the discovery of chloroform. Syme, however, opposed the new idea. Acupressure finally failed, partly due to the antisepsis introduced by Lister which was to prove the major factor in the prevention of wound infection. After the hard work Simpson had put into its development, the failure must have been a severe disappointment.

The scope of surgery in the nineteenth century was strictly limited. Operations on the large ovarian cysts which developed in some unfortunate patients were rare and hazardous. Simpson encouraged the early pioneers, though Syme considered their efforts fraught with danger. It was in such a case, the surgical treatment of an ovarian mass, that Simpson experienced the death from chloroform previously referred to.

One of the last subjects to interest Simpson in his lifetime was hospitalism. He had earlier noted that more fatal cases of puerperal sepsis occurred when patients were delivered in hospital than when they had their babies at home. He came to believe in the

Joseph Lister, whose introduction of antisepsis made acupressure unnecessary.

dangers of large hospitals and the advantages of treating patients in small isolated units, with provision for their removal and rapid reconstruction in an effort to combat infectious outbreaks. Another man with similar ideas was the great engineer Isambard Brunel, who planned a prefabricated hospital for use in the Crimean war. It had planned ventilation systems and advanced washing and toilet facilities, but was never used to full capacity.

Simpson, however, has never been given credit for these ideas. Circumstances contrived to rob him of success, while the principles advocated by Lister gained ground and displaced Simpson's arguments in the minds of the leading surgeons of the day. Lister was son-in-law to Syme, and it appears ironic that his ideas were accepted somewhat tardily by his own father-in-law and never made a great impression on the mind of Simpson.

In February 1870 Simpson journeyed to London to give evidence in a celebrated divorce case. On the journey back to Edinburgh he suffered pain in the chest which was severe enough to cause him to lie on the floor of the railway carriage. He carried on with his work, however, for a time. Then the chest pain returned and he was forced to his bed. It seems that this was the pain of angina pectoris or coronary insufficiency, the pain developing when the heart muscle fibres have to contract with an inadequate supply of oxygen. In bed he read and received visitors. In April he had the bed moved downstairs to the back drawing-room on the first floor. Gradually his condition worsened and everybody realized that death was near. The family gathered and on 5th May his elder brother, Sandy, sat on the pillow, James's head on his knee. On the morning of the 6th he was unconscious and he died that day.

The public had followed the last illness with interest and sorrow. The funeral took place on Friday, 13th May, with the flags of the city at half

Simpson's grave in Warriston cemetery.

mast. A vast crowd followed the procession down Queen Street to Warriston Cemetery, where he was laid to rest in the family plot.

Letters of condolence reached 52 Queen Street from many quarters and included a message from Her Majesty. There had been a move to hold a state funeral in Westminster Abbey, but the family preferred the Edinburgh burial. It was, however, decided to erect a monument and a statue to the great obstetrician in Edinburgh, and a marble bust in Westminster Abbey.

Jessie did not long outlive her husband. She died on 17th June at Killin and was brought back to Edinburgh for burial in the family plot at Warriston with her husband and children.

Simpson's statue in Princes Street, Edinburgh.

8 Edinburgh Today

A hundred years after Simpson's death his memory is still revered. In Edinburgh a hospital carries his name and his statue looks out over Princes Street. The Warriston Cemetery contains memorials to members of the family buried there.

Fifty-two Queen Street passed into the keeping of James's nephew, Alexander Simpson, who also succeeded to the Chair of Midwifery. In 1916 the house came into the ownership of the Church of Scotland. In 1947 the Queen opened the house as a youth leadership training centre. Recently it has been acquired by the Committee on Moral Welfare of the Church of Scotland as a centre for the care of alcoholics and drug addicts. The house was once more opened by this Committee on 6th May 1970, as the Mound Centre. The date was the centenary of Simpson's death, and many of his family were present at the ceremony.

The Centre has preserved Simpson's front ground floor dining room as the "Discovery Room" in the style of when the family lived in the house. In the room are kept some of the furniture, a clock, porcelain figures, vegetable dishes and personalia. These latter include a cross which doubles as a lancet for the bleeding of patients and a pillbox. The first floor drawing room is now a lounge cafeteria, while the back room in which Simpson died has been converted to a chapel. The rear ground floor room where Simpson kept his books and conducted prayers daily is now a classroom and demonstration room for students. The Mound Centre is surely a project which the spirit of Simpson would approve. As a young man he had been interested in sociological as well as medical problems in Edinburgh and it is fitting that the house is used today to help in the care of those in need.

Date Chart

1811   Birth of James Simpson, 7th June.
        Baptism, 30th June.
1820   Death of Mary Simpson, James's mother.
1825   Began studies at the University of Edinburgh.
1830   Death of David Simpson, James's father.
        Qualification as Member of Royal College of Surgeons.
1831   Chloroform prepared by von Leibig, Guthrie, Soubeiran (independently).
1832   MD qualification obtained.
        Assistant to Professor of Pathology.
1835   Dumas described the properties of chloroform.
        Visited Liverpool and met Jessie Grindlay.
1836   Deputy Professor of Pathology.
1838   Lecturer in Midwifery.
1839   Application for Chair of Midwifery.
        Married Jessie Grindlay, 26th December.
1840   Elected to the Chair of Midwifery.
1845   Moved to 52 Queen Street.
1846   Morton's demonstration of ether, Boston, Massachussetts, U.S.A., 16th October.
1847   Simpson used ether in obstetrics, 19th January.
        Inhalation of chloroform, 4th November.
        Read paper on chloroform to Edinburgh Medico-Chirurgical Society, 10th November.
        Classic paper published in *The Lancet*, 20th November.
        Appointed Physician Accoucher to the Queen for Scotland.
1853   John Snow administered chloroform to Queen Victoria.
1861   President of the Society of Antiquaries of Scotland.
1866   Baronetcy.
        Doctorate of Civil Law, University of Oxford.
1870   Died, 6th May. Buried Warriston Cemetery.

Further Reading

Duns, J., *Memoir of Sir James Y. Simpson, Bart.* (Edmonston and Douglas, Edinburgh, 1873).

Richardson, Sir B. W., *John Snow, M.D., A Representative of Medical Science and Art of the Victorian Era* (The Asclepiad, London, 1887, Vol. IV, pp. 274–300).

Laing Gordon, H., *Sir James Young Simpson and Chloroform* (T. Fisher Unwin, London, 1897).

The History of Duncan Flockhart & Co. Commemorating the Centenaries of Ether and Chloroform (Duncan, Flockhart & Co., Edinburgh and London, 1946).

Duncum, Barbara M., *The Development of Inhalation Anaesthesia* (Oxford University Press, London, New York, Toronto, 1947).

Sykes, W. S., *Essays on the First Hundred Years of Anaesthesia* (E. and S. Livingstone Ltd., Edinburgh, Vol. 1 1960, Vol. 2, 1961).

Rowland, J., *The Chloroform Man. The Story of Sir James Simpson* (Lutterworth Press, London, 1961).

Keys, T. E., *The History of Surgical Anaesthesia* (Dover Publications Inc., New York Constable & Co. Ltd., London, 1963).

Armstrong Davison, M. H., *The Evolution of Anaesthesia* (Sherratt, Altrincham, 1965).

Faulconer, A. and Keys, T. E., *Foundations of Anesthesiology* (Thomas, Springfield, Ill., Vols. 1 and 2, 1965).

Baillie, T. W., *From Boston to Dumfries. The First Surgical Use of Anaesthetic Ether in the Old World* (Dinwiddie, Dumfries, 1966).

MacQuitty, Betty, *The Battle for Oblivion. The Discovery of Anaesthesia* (Harrap, London, Toronto, Wellington, Sydney, 1969).

Shepherd, J. A., *Simpson and Syme of Edinburgh* (E. and S. Livingstone, Ltd., Edinburgh and London, 1969).

Progress in Anaesthesiology—Proceedings of the Fourth World Congress of Anaesthesiologists, London 1968 (Excerpta Medica Foundation, Amsterdam, 1970, pp. 164–232).

Simpson, Myrtle, *Simpson the Obstetrician* (Victor Gollancz Ltd., London, 1972).

Glossary

ACUPRESSURE   A method for arresting haemorrhage during surgery.

ADRENALINE   A substance produced in the adrenal glands of the body, which causes constriction of blood vessels and stimulates the heart.

ANATOMY   The study of the structure of the body.

ANGINA PECTORIS   Chest pain caused by the heart beating with inadequate oxygen.

ANTISEPSIS   The removal of living organisms from the surgical field, to prevent infection of the wound.

CLINICIAN   A physician or surgeon treating patients.

CRANIOTOMY   An obstetric operation for delivery of the head, when the baby is already dead.

FEMUR   The long bone in the thigh.

HOMOEOPATHY   Treatment of disease by administration of small amounts of drugs which mimic that disease.

HOSPITALISM   The study of the planning and function of hospitals.

LANCET   Surgical knife.

MIDWIFERY   The art of childbirth.

OBSTETRICS   The science of midwifery.

OBSTRUCTED LABOUR   Complication of labour, due to obstruction to delivery of the baby.

PATHOLOGY   The study of disease.

PUEPERAL SEPSIS   Infection of the womb following childbirth.

SEPTICAEMIA   Infection of the blood.

SYMPATHETIC NERVOUS SYSTEM   A part of the nervous system which operates without consciousness. Associated with the action of adrenaline.

VAGAL INHIBITION   Stimulation of the vagus nerve to the heart, causing cessation of the heart beat.

VENTRICULAR FIBRILLATION   Irregular, uncoordinated contraction of the heart muscle fibres, so that the heart beat is ineffectual.

Index

Abbott, Gilbert, 29, 32
Acupressure, 85
America, 29, 31, 33, 34, 35, 41, 47, 59, 60, 73
Amputations, 18, 34–35, 54, 63
Anatomy, 16, 63
Anaesthesia, 18, 29, 31, 33, 35, 39, 46, 47, 50, 54, 55, 56, 59, 63, 68, 73, 74, 76
—, chloroform, 15, 31, 42, 45, 46, 47, 50, 52, 54, 55, 56, 57, 59, 60, 63, 67, 68, 71, 73, 74, 76, 81, 85
—, ether, 18, 31, 32, 33, 35, 39, 41, 45, 46, 47, 54, 56, 59, 68
nitrous oxide, 59
Australia, 39

Bathgate, 11, 12, 15, 16, 17, 19, 25, 26
Bigelow, Dr. H. J., 29, 33
Boot, Dr., 34, 35
Britain, 18, 29, 31, 39, 40, 50, 54, 60, 73
Burke, William, 16–17

Carbon dioxide, 30
Channing, Dr. W., 41, 81
Childbirth, 39, 45, 74, 76; *see also* midwifery, obstetrics

Chloric ether, 72
Chloroform *see* Anaesthesia, chloroform
Clarke, W. E., 31
Clover, Joseph, 59
Colton, G. Q., 31, 59
Coote, Holmes, 72
Crimean War, the, 50, 86

Davy, Humphrey, 29
Dublin, 26, 27, 73
Dumas, Professor, 42, 46, 81
Duncan & Flockhart Co., 42, 49–50, 52, 53
Duncan, John, 49–50, 53
Duncan, Matthews, 42, 45

Edinburgh, 11, 12, 15, 17, 18, 19, 21, 25, 26, 34, 35, 39, 41, 46, 47, 49, 50, 55, 63, 67, 72, 84, 86, 87, 89
—, *Edinburgh Medical Journal*, 24
—, Royal College of Surgeons, 19, 21, 35, 49, 50
—, Town Council, 26, 27, 84
—, University, 15, 22, 35, 63, 84
Elliotson, John, 31
Esdaile, John, 31
Ether *see* Anaesthesia, ether

Flockhart, Mr., 50, 53
Forbes, Edward, 22
France, 11, 22, 24, 41, 42
Fraser, Dr., 35

Germany, 24, 41, 80
Glasgow, 11, 12, 24, 35
Greener, Hannah, 54–55
Grindlay, Jessie, *see* Simpson, Jessie (wife)
—, Walter, 22, 24, 26, 27

Halothane, 59–61
Hamilton, James, 18, 26
Hare, William, 16–17
Henderson, William, 65
Homoeopathy, 65
Hospitalism, 85–86
Hypnotism, 30

India, 31, 47
Inhalations, 41, 42, 45
Ireland, 26, 84

Jackson, Dr. Charles T., 31, 47
Jarvie, George, 15
—, Mary, 11

Keith, Dr. George, 42, 45
Kennedy, Dr. Evory, 26, 27

Knox, Robert, 16–17, 21

Lancet, The, 29–30, 35, 45
"Laughing gas" see nitrous oxide
Lister, Joseph, 85, 86
Liston, Robert, 18, 34–35, 39, 54, 63
Liverpool, 22, 25, 26, 27, 33, 35, 42, 49
London, 11, 17, 22, 24, 35, 39, 47, 49, 54, 55, 59, 66, 67, 72, 73, 82, 86
Long, Crawford Williamson, 31

MacArthur, Mr., 16
MacLagan, Sir Douglas, 22
Midwifery, 18, 21, 25, 26, 41, 65, 74, 85, 89
Miller, Professor, 41, 46, 66, 67
Morton, William, 29, 31–33, 47

Nitrous oxide, 29, 31, 59; see also Anaesthesia, nitrous oxide

Obstetrics, 18, 25, 26, 41, 45, 73, 81

Paris 22, 46, 54, 59
Pathology, 21, 22, 25, 65
Pugh, Dr. W. R., 39

Reid, John, 15, 16, 17, 25
Royal College of Surgeons, 19, 21, 35, 49, 50
Royal College of Surgeons of England, 68
Royal Medical Society, 22, 24, 57

Scott, Dr., 35
Scott, Sir Walter, 22
Simpson, Alexander (grandfather), 11
—, Alexander (Sandy) (brother), 21, 25, 26, 53, 86
—, children of, 79–81
—, David (father), 11, 12, 19
—, David (brother), 21
—, Isabella (grandmother), 11
—, James, 11, 12, 15, 16, 17, 18, 21, 22, 24, 25, 26, 27, 35, 39, 41, 42, 45, 46, 49, 50, 52, 53, 54, 55, 56, 63, 65, 66, 67, 68, 72, 73, 76, 79, 80, 81, 82, 84, 85, 86, 89
—, publications of: Answer to Religious objections advanced against the employment of Anaesthetic agents in midwifery and surgery, 74
—, "Diseases of the Placenta", 24

—, "On a New Anaesthetic Agent", 45
—, Jessie (wife), 11, 24, 25, 26, 27, 66, 67, 79, 87
—, John (brother), 22
—, Mary (mother), 11, 12, 15
—, Mary (sister), 15, 24, 25, 26
Slavery, 22, 29
Snow, John, 55, 67, 68, 71, 72, 73, 74, 76
St. Andrews University, 16
Surgery, 18, 22, 29, 30, 33, 35, 45, 46, 54, 60, 63, 65, 66, 71, 72, 74, 85
Syme, James, 63, 65, 66, 67, 72, 73, 82, 85, 86

Thomson, Dr. John, 21, 22, 25

University College Hospital, 18, 35, 63

Vagal inhibition, 56
Ventricular fibrillation, 56
Victoria, Queen, 49, 71, 74, 76, 80, 87

Wakley, Thomas, 29
Waldie, David, 42, 49
Warren, Dr., 33
Wells, Horace, 31
Wells, Spencer, 81

Appendices

Part of the Family Tree

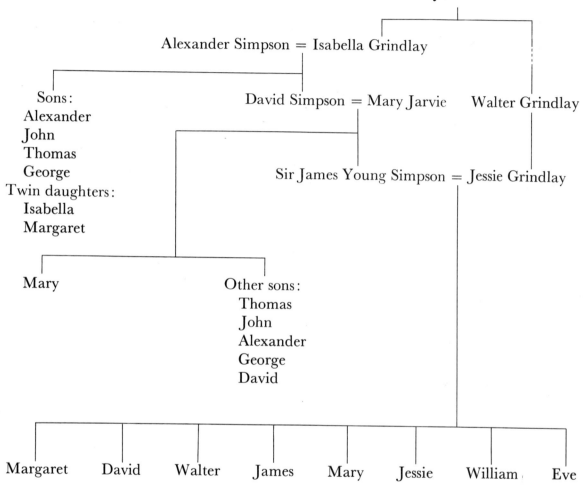

Chemical Formulae

N_2O NITROUS OXIDE

$C_2H_5-O_2-C_2H_5$ ETHER

 CHLOROFORM

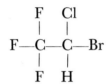 HALOTHANE

Professors at Edinburgh University

PROFESORS OF MIDWIFERY

1726	Joseph Gibson
1739	Robert Smith
1756	Thomas Young
1780	Alexander Hamilton
1800	James Hamilton
1840	Sir J. Y. Simpson
1870	Alexander Simpson

PROFESSORS OF CLINICAL SURGERY

1803	James Russell
1833	James Syme
1869	Joseph Lister

PROFESSORS OF SURGERY

1836	Sir Charles Bell
1842	James Miller